Iron Ores

Edited by Volodymyr Shatokha

Published in London, United Kingdom

IntechOpen

Supporting open minds since 2005

Iron Ores
http://dx.doi.org/10.5772/intechopen.87814
Edited by Volodymyr Shatokha

Contributors
Alla A. Novakova, Dmitrii S. Novikov, Hanna Krztoń, Janusz Stecko, Jaroslav Legemza, Mária Fröhlichová, Róbert Findorák, Martina Džupková, Yıldırım İsmail Tosun, Sumant Samal, Manoja Kumar Mohanty, Subash Chandra Mishra, Bhagiratha Mishra, Volodymyr Shatokha, Mykola Stupnyk, Haisheng Han, Wenjuan Sun, Wei Sun, Yuehua Hu

First published in London, United Kingdom, 2021 by IntechOpen
IntechOpen is the global imprint of INTECHOPEN LIMITED, registered in England and Wales, registration number: 11086078, 5 Princes Gate Court, London, SW7 2QJ, United Kingdom
Printed in Croatia

British Library Cataloguing-in-Publication Data
A catalogue record for this book is available from the British Library

Additional hard and PDF copies can be obtained from orders@intechopen.com

Iron Ores
Edited by Volodymyr Shatokha
p. cm.
Print ISBN 978-1-83962-550-3
Online ISBN 978-1-83962-551-0
eBook (PDF) ISBN 978-1-83962-558-9

For EU product safety concerns:
IN TECH d.o.o., Prolaz Marije Krucifikse Kozulić 3, 51000 Rijeka, Croatia, info@intechopen.com or visit our website at intechopen.com.

Meet the editor

Volodymyr Shatokha graduated in 1982 as a Ferrous Metallurgy engineer. In 1985, he received a Ph.D. degree from the Dnipropetrovsk Metallurgical Institute (known since 1999 as the National Metallurgical Academy of Ukraine). In 1998, he received the title of professor with the Ironmaking Chair. During 1999–2002, he was the dean of the Metallurgical Faculty. Since 2002, he has worked as the vice-rector for Research and Education. His research deals mainly with the properties of iron ore materials, physicochemical processes of ironmaking, recycling methods, sustainability analysis, etc. He is the author of over 180 research papers, 7 books, and 7 patents. He is the honorary professor at the Inner Mongolia University of Science and Technology, China, and a visiting professor at the University of Tokyo (2013). He is a laureate of the State Prize of Ukraine in Science and Technology (2010) and a fellow of the Erasmus+ Jean Monnet Programme for the EU climate and energy policy studies (since 2015).

Contents

Preface

During the two decades of the 21st century, the yearly steel output grew more substantially than within the entire previous lifetime of human civilization and so did the iron ore mining. Thus, despite all the progress we made in the delivery of novel nonferrous materials, we continue to live in the Iron Age with steel, an alloy of iron and carbon, being the most important engineering and construction material worldwide. Taking into account the very well-known advantages of steel, including its nearly 100% recyclability and abundance of iron-ore resources, the importance of mining and processing iron ores will not vanish for the years to come.

This book covers research areas encompassing the value chain from mining to processing of iron ore and is structured into two sections: (i) Genesis and Exploitation and (ii) Characterisation and Processing. The chapters are written by researchers from all over the world covering the multidisciplinary aspects from geological genesis to historic context of mining and exploitation of deposit and from beneficiation and sintering of iron ore to novel reduction techniques and application of advanced characterization methods.

I am very grateful to all the authors for their valuable contributions, willingness to share the research results, and expertise with the reading audience and for their hard work to make this book happen.

I sincerely thank IntechOpen for the initiative to publish this title. The readers-researchers, students, and industry professionals, now have a rare opportunity to find comprehensive, cutting-edge information covering a wide range of interdisciplinary issues in the domain of science and technology of iron ores in a single book.

Volodymyr Shatokha
National Metallurgical Academy of Ukraine,
Dnipro, Ukraine

Section 1

Genesis and Exploitation

Chapter 1

Study of Deep-Ocean Ferromanganese Crusts Ore Components

Alla A. Novakova and Dmitrii S. Novikov

Abstract

A complex layer-by-layer morphology and phase analysis of a ferromanganese crust aged about 70 million years, extracted from the rise of the Magellan Mountains of the Pacific Ocean, was carried out using several physics methods: digital optical microscopy, scanning electron microscopy with high resolution, X-ray fluorescence and diffraction analysis and Mossbauer spectroscopy. This analysis showed that the crust is an association of several minerals with various dispersion and crystallization degree, between which fossilized bacterial mats with Fe- and Mn- oxides are located. These phenomena indicate the biogenic nature of the crust. Changes in the crusts phase composition from the lower layer to the upper layer indicate changes in the external environmental conditions during their formation.

Keywords: Ferromanganese Crusts, X-ray analysis, Scanning electron microscopy, Mössbauer spectroscopy, Bacterial activity

1. Introduction

Ferromanganese crusts were first discovered at the bottom of the Pacific Ocean. The crusts samples were taken on board by the English ship "Challenger" and described in 1873 during the first complex oceanological expedition in the history of world science. However, until the middle of the XX century, only the chemical composition of crusts was analyzed. It was found out that ferromanganese crusts consist of up to 20% Mn, 15% Fe and 0.3–2.5% Co. In addition, they can contain complexes of noble, non-ferrous, rare and rare earth metals (up to 70 elements). All this data obtained stimulated further complex investigations on the topic. Many expeditions, mainly in the Pacific Ocean, were conducted to find and obtain ferromanganese crusts by USSR, USA, Germany, China, Japan and other countries in 1970–1980. Currently, crusts are found in all oceans, including the Arctic Ocean. They are often spread on basalts and clastic volcanic rocks in the depth range of 400–7000 m. The crusts are relatively thin continuous covers lying on the slopes of seamounts of volcanic origin. The thickness of the crusts can reach 26–40 cm. The crusts age is usually determined by Be, U, and Os – isotope dating methods [1]. The average age reaches several tens of millions of years. The most common crustal growth rates found in all oceans vary between 2 and 4 mm/mln.yr.

Researchers of ferromanganese crusts have always had 2 questions: 1) where do the colossal masses of manganese and iron come from to the places where the crusts form? and 2) what is the mechanism of the crusts formation and why does it differ a lot from the usual bottom depositions? Until the middle of the XX century, the idea of hydrogenic mineralization (the deposition of metals from the oceanic water) prevailed. It is known that the flow of substances from various sources into the ocean water is dynamically balanced by their removal to the bottom depositions. Thus, the salt composition of ocean water remains stable. However, model experiments and real observations did not reveal the deposition of large amounts of iron and manganese to the ocean floor. In the second half of the XX century, more attention was paid to the underwater volcanic eruptions, mantle fluid flows and post-volcanic hydrothermal activity i.e. processes that are actually recorded, which may be the main suppliers of metals for the oxide ferromanganese ores formation. Accordingly, the volcanogenic-sedimentary type of mineralization was identified.

By the end of the 1990s, marine geologists, microbiologists, and micropaleontologists had developed and validated the biological concept of ferromanganese oxide ore formation. According to this concept, crusts are considered as products of the vital activity of bacterial communities that can oxidize divalent iron and manganese compounds and precipitate metal oxides in a crystalline or amorphous form on the cell surface or even within the cell, as well as in the matrix of biological films [2, 3]. The biological concept of oxide ferromanganese ore genesis has been brilliantly confirmed by scanning electron microscopy [4].

Developing and fossilizing biofilms (0.5 to 2 microns thick or more) form bacterial mats with a multilayer structure: stand-alone bundles of biofilms separated by cavities, clusters of filamentous bacteria, and layers of glycocalyx. Therefore, in bacterial mats, dense, massive micro-layers and porous, loose ones are distinguished. It is bacterial mats that form columnar stromatolites of ferromanganese crusts, which are the ore components. During the entire time of stromatolite growth (millions of years), extreme events periodically occurred (underwater volcanic eruptions, tectonic phenomena, global glaciation, etc.) that affected the vital activity of microorganisms and were imprinted in the crustal sections by the formation of interlayers with different types of columns (the thickness and density of columns as well as their growth direction changed, and bushy branches were formed) [5]. At the same time, the growing crusts, having a high porosity and a fine structure, demonstrate a large sorption capacity. As a result, they are saturated with complexes of non-ferrous, rare and rare earth metals. The biofilm matrix contains a significant concentration of polysaccharides with a negative charge. Due to this, metal cations are able to accumulate on the surface of the extracellular polymer matrix, forming strong complexes. These processes can explain the mechanism of enrichment of ferromanganese crusts with the ore-compound metals [6].

The structure, composition and genesis of ferromanganese crusts have been studied by scientific laboratories in many countries for more than 50 years (since their industrial significance was discovered). However, the nature of the oxide ferromanganese ore formation has not been fully revealed yet. This is due to the fact that the crusts are very complex in composition and structure layered formations. The nanoscale oxide ore components of biogenic origin are also the significant part of the crusts composition in addition to numerous clastic and dispersed minerals formed as a result of volcanic activity. Therefore, it is obvious that the study of such objects requires subtle chemical and physical methods of studying their composition, structure and morphology, which appeared only at the end of the XX century. This work is devoted to such research.

2. Analysis of ferromanganese crusts phase composition and morphology

A characteristic feature of ferromanganese crusts is their layered structure, which shows a history of their origin and growth. If the crust is sliced, one can usually observe from 3 to 5 layers with an average thickness of 2–3 cm, differing in structure, physical properties and composition of components. Usually, the lower layers of the crust are dense and strong, while the upper layers are more porous and brittle. It is possible to consistently trace the processes of ore accumulation and global changes in the external conditions for the formation of ferromanganese crusts by studying the composition and structure of these layers.

2.1 Object of study

Figure 1 shows a section of the crust studied in our work, isolated at a depth of 1200 m from the surface of the guyot of the Magellan Mountains of the Pacific Ocean. This crust has a special feature: it has a "relic" layer (R) of pre-existed crusts that underlies the main section and situated on the weathered basalt. The thickness of the R – layer reaches 8 cm. The stratification of the crust section was carried out by marine geologists M. Melnikov and S. Pletnev (from the Institute of Oceanology, Gelendzhik). The layers are named (below): I-1, I-2, II and III, which were formed in the time intervals indicated on the geochronological scale placed to the left of the crust.

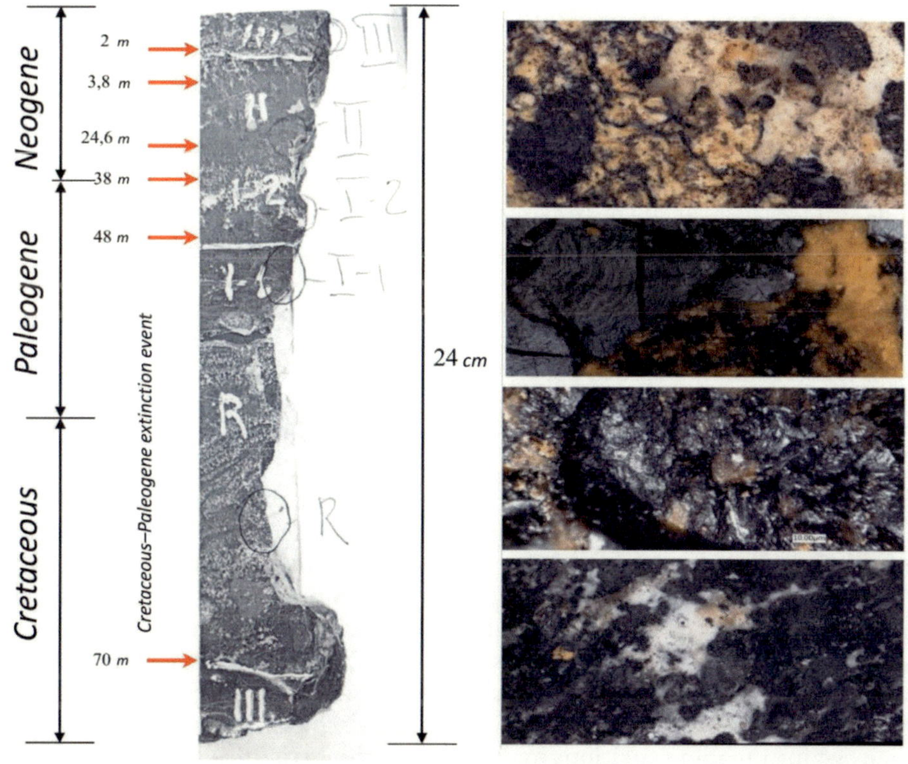

Figure 1.
Section of the studied sample of the ferromanganese crust, compared with the geochronological scale (in the interval of 70 million years). Optical images of the various layers are on the right.

The method of layer-by-layer studies of the crust was as follows:

1. Color optical images were obtained from all layers of the crust section using the Keyence VHX – 5000 digital optical microscope which allows us to record changes in the phase composition of the crust from layer to layer.

2. A TESCAN VEGA scanning electron microscope was used to study the morphology and relative position of various crystal phases, as well as to search for traces of bacterial activity that leads to the growth of ore formations

3. The elemental composition of each layer at individual points along its surface was determined using a portable X-ray fluorescence analyzer SciAps X-200, which has a collimated X-ray beam with a diameter of less than 3 mm. These results were then mathematically processed to obtain the average elemental composition for each layer of the crust.

4. The results of the elemental composition were entered into the High Score Plus program of the PANalytical Empyrean X-ray diffractometer, which was used for X-ray diffraction analysis to determine the nature of ore and accessory minerals, and assess accompanying growth of ore formations.

5. The Mössbauer spectra were taken and analyzed for each layer to clarify the phase composition of the ultrafine iron-containing compounds that make up the thoracic component. Mathematical processing of the obtained Mossbauer spectra made it possible not only to determine which iron oxide compounds are contained in each layer of the crust, but also to trace quantitative and qualitative changes in their composition during the transition from one layer to another.

2.2 Experimental microscopy results

The surface layers of the ferromanganese crust analysis performed by digital optical microscopy (**Figure 1**) showed that they consist of associations of a large number of minerals with different degrees of crystallization and dispersion. There is also an increase in yellowish-red inclusions from the lower layer I to the upper layer III. These inclusions correspond to iron oxide compounds (goethite, hematite, etc.), i.e., there is an increase in the iron-ore component from the lower layer of the crust to the upper one, and the proportion of accessory minerals decreases. It is difficult to analyze the content of the manganese-ore component in the crust layers by this method, since manganese oxides are very dark, and it is hard to distinguish them from other dark-colored phases. However, we were able to identify an interesting region that has a colomorphic structure, which is most likely formed by manganese oxides (**Figure 2a**) by examining the sample from the R-layer in detail [7].

We examined this region of the R-layer using a scanning electron microscope (**Figure 2b**) to clarify the assumption about the manganese nature of the colomorphic structure. The Energy Dispersive analysis performed on the electron microscope allowed us to obtain a picture of the distribution of chemical elements in this region of the R-layer (**Figure 3**). The analysis shows that manganese and oxygen occupy the same positions in the colomorphic structure, which indicates that they are combined in the form of manganese oxide. At the same time, calcium and phosphorus also occupy identical positions, which indicates that they are connected in a common structure, most likely apatite. No iron compounds were found in this area.

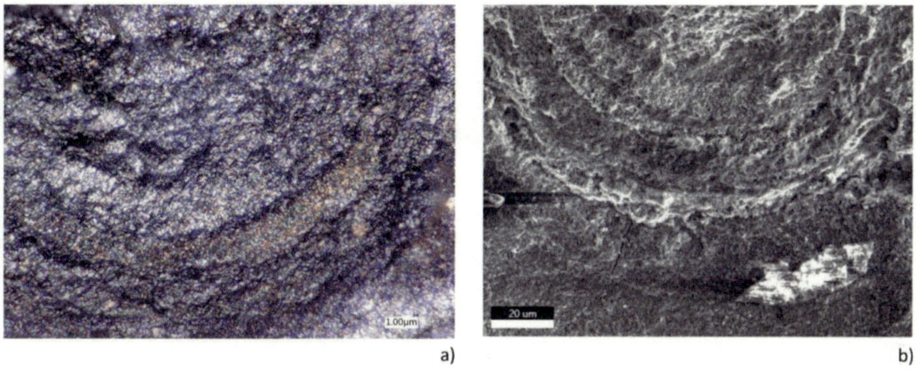

Figure 2.
Optical (a) image of a colomorphic structure on the surface of the R-layer and electron microscopic image (b) of the same structure.

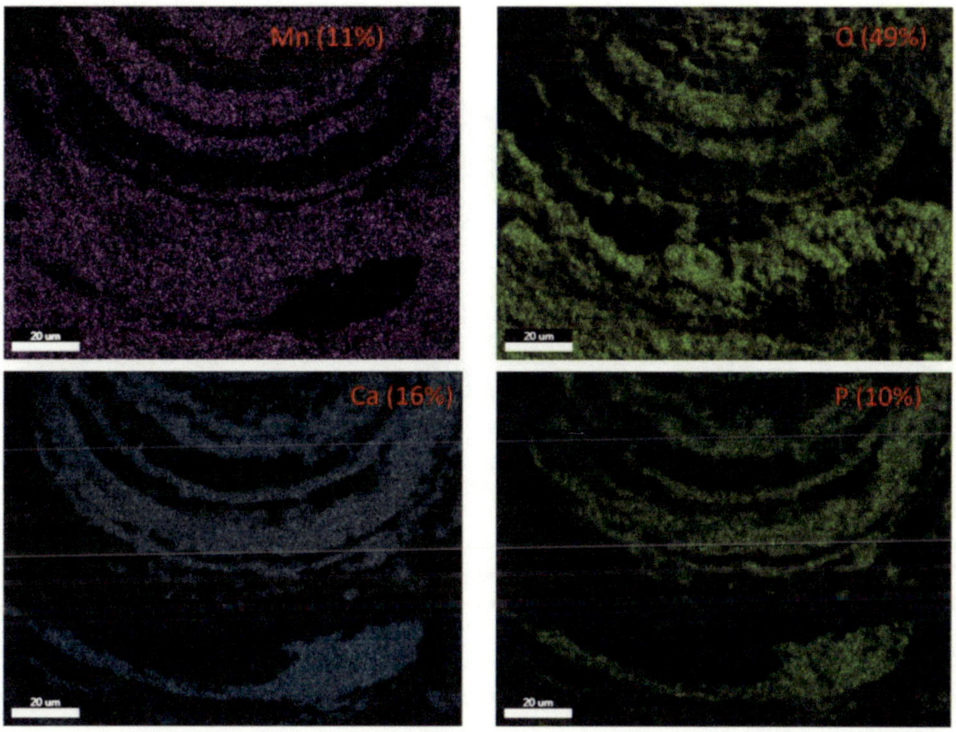

Figure 3.
Distribution of chemical elements in a colomorphic structure on the surface of the P-layer and their percentage.

By increasing the magnification of the scanning microscope (x10, 000), we were able to see columns of stromatolites, layers of bacterial mats with fossilized bacteria, and microcrystals of accessory minerals (**Figures 4** and 5).

2.3 X-ray investigation results

Elemental analysis of each layer of the studied crust was carried out at 8 separate points of its surface area (a surface area was selected, which will later be analyzed using an X-ray diffractometer) using the SciAps X-200 analyzer. The built-in camera made it possible to obtain an image of the analyzed area of the sample and accurately determine the measurement point of the X-ray spectrum.

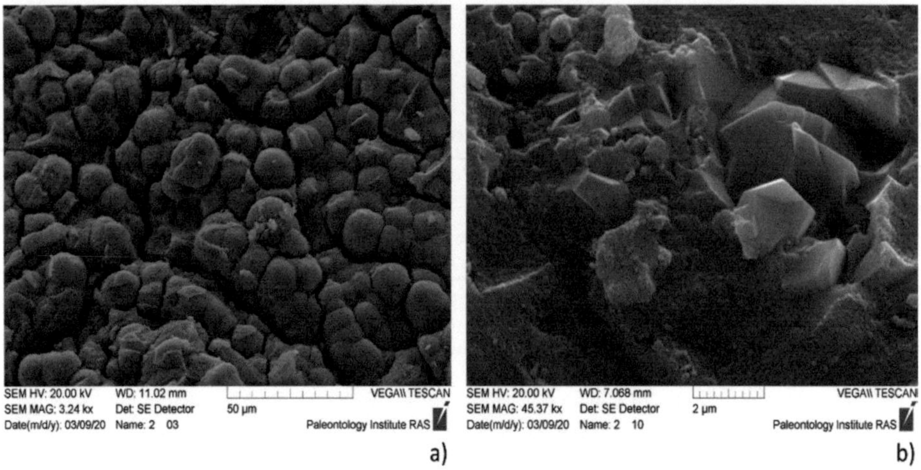

Figure 4.
Micrographs of the ferromanganese crust with different magnifications: (a) numerous columns of stromatolites are seen (top view); (b) microcrystals of apatite lying on bacterial mats.

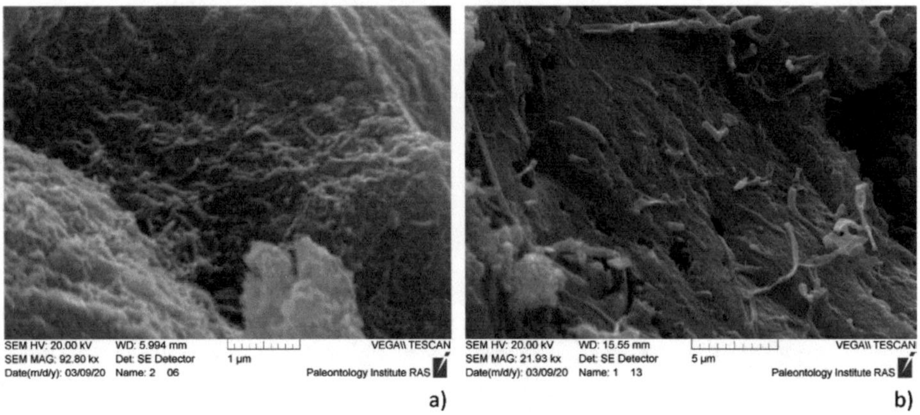

Figure 5.
Micrographs of bacterial mats: (a) a mat completely covered with filamentous bacteria; (b) a mat with void zones formed by cyanobacteria.

Mathematical processing of the obtained data allowed us to find out the elemental composition of each layer under study, the numerical data of which are presented in **Table 1**.

According to **Table 1**, up to 20 different elements – metals, as well as phosphorus and silicon-were found in each layer of the studied crust. The average mass ratio in the crust is dominated by Fe (%), Mn (%), and P (%). Si and Al are also present in significant amounts. All these data were used to conduct a layer-by-layer phase analysis of the studied crust.

Figures 6–8 shows X-ray diffractograms of all layers of the studied crust. The diffractograms were decoded for all phases, but we reverse special attention (highlight in the diffractograms) on Fe and Mn oxides. The main mineral phase in the relict layer (**Figure 6a**) is Fluorapatite, as well as minerals (calcium silicate, corundum, pargasite and pyrosmalite). Fe and Mn oxides compounds are detected in **Figure 6b**.

It is extremely difficult to perform a quantitative phase X-ray analysis of the studied layers of the ferromanganese crust: they consist of associations of a large number of minerals and nanoscale, often amorphous, oxide compounds of iron and manganese.

Sample	R	I-1	I-2	II	III
Al	3.78	1.02	2.17	4.33	3.49
Bi	0.11	0.09	0.00	0.00	0.00
Co	0.28	0.00	0.00	0.00	0.00
Cr	0.00	0.07	0.04	0.13	0.00
Cu	1.64	0.86	0.81	0.75	0.76
Fe	13.06	40.17	37.33	25.74	45.11
Mg	2.15	0.66	1.12	1.00	1.58
Mn	44.81	46.53	33.47	20.64	30.63
Mo	0.11	0.26	0.10	0.03	0.12
Nb	0.03	0.04	0.03	0.07	0.02
Ni	5.16	2.27	1.38	2.06	1.42
P	36.44	0.00	25.82	26.69	9.70
Pb	0.27	1.21	0.62	0.28	0.44
S	0.00	0.00	0.00	1.66	0.00
Sb	0.00	0.12	0.05	0.08	0.00
Si	2.68	2.94	3.97	11.00	8.63
Sn	0.00	0.09	0.00	0.08	0.04
Ti	2.75	2.08	1.79	3.00	1.44
V	0.88	0.37	0.31	0.49	0.21
W	0.08	0.07	0.08	0.09	0.07
Zn	0.91	0.65	0.43	0.48	0.42
Zr	0.46	0.78	0.68	0.74	0.60

The lines of the main elements of the ore component Fe and Mn are highlighted in color for clarity of their change from layer to layer.

Table 1.
Elemental composition obtained by means of X-ray fluorescence analysis.

As a result, the X-ray reflexes of oxides are very weakened and broadened in comparison with the reflexes of minerals of microcrystals. Their intensity is often comparable to the error limit of the method (2%). Therefore, we performed only to qualitative analysis.

The observed broad peaks of Mn- oxides were deciphered by introducing todoroctitis and unstable buserite, which explains significant increase in the intensity and area of the peaks in the previously identified diffractogram angles: 35° - 37° and 63° - 67°. Iron oxides on diffraction patterns (**Figures 6–8**) are represented by goethite and hematite.

2.4 Mossbauer spectroscopy

More accurate studies of nanoscale iron oxide compounds in different layers of CMC can be carried out by the method of Mössbauer spectroscopy on the Co^{57} isotope. Due to its selectivity (only the Fe^{57} nuclei reflexes are recorded in the spectra), the method makes it possible to separate iron-containing minerals from the total mineral mass. According to the parameters of the Mössbauer spectrum. It is possible to determine the valence state of iron, the symmetry of its local neighborhood from the Mössbauer spectra parameters and therefore to determine the iron-containing

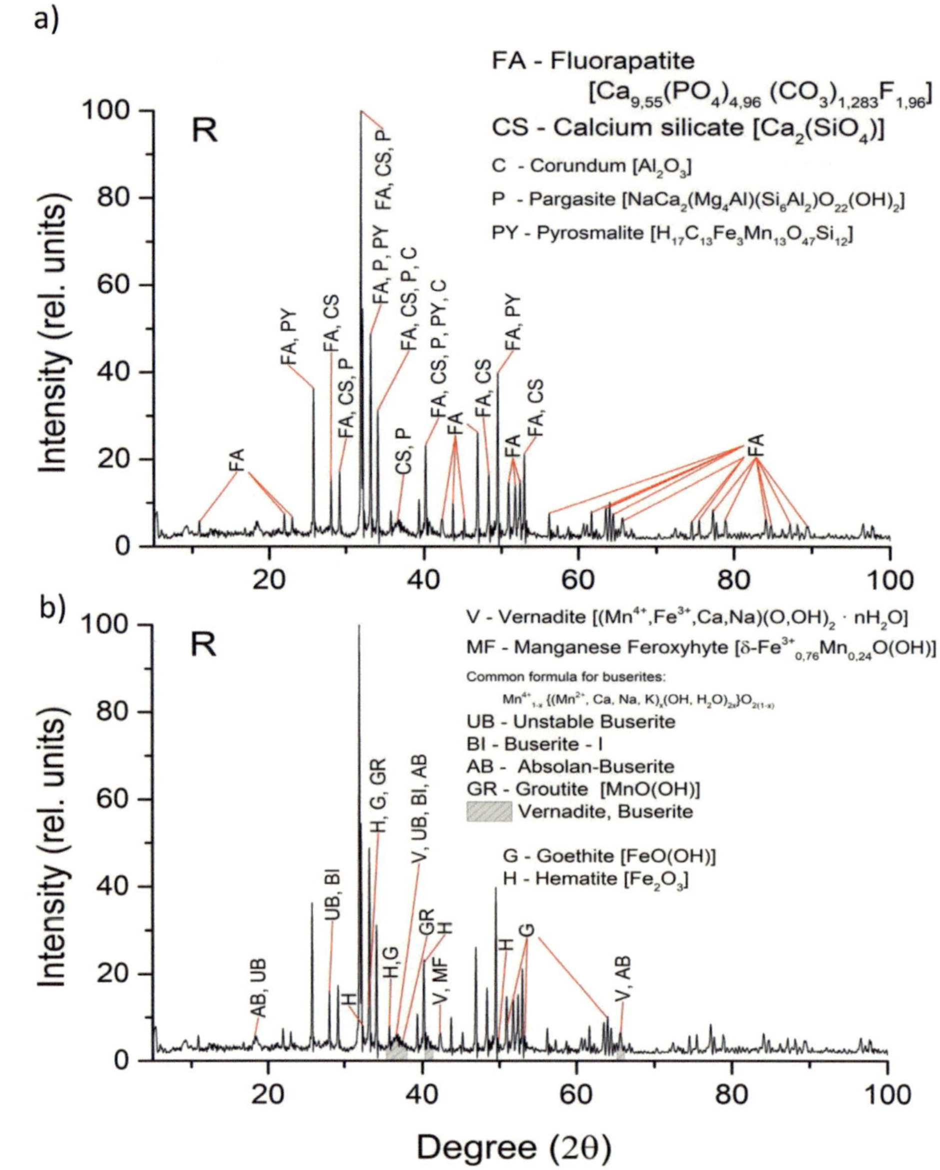

Figure 6.
X-ray diffractogram of the relic layer of the ferromanganese crust: (a) analysis of minerals components; (b) analysis of Fe- and Mn- oxides.

mineral [8], as well as to estimate the particle sizes of iron compounds from the temperature dependence of Mössbauer spectra [9, 10].

We performed the Mossbauer investigation of all the Iron-manganese crust layers (**Figure 1**) at room temperature. Five spectra of the different layers are presented in **Figure 4**.

All the spectra are broadened paramagnetic doublets, which indicates the superparamagnetic state of the iron oxide particles [11]. Mathematical processing of the spectra made it possible to decompose them into components corresponding to various iron oxides. Thus, a layer-by-layer quantitative phase analysis of the iron compounds of the studied crust was carried out. The figure shows that the smallest effect value (4.6%) is observed in the relic layer. There are significantly fewer iron-containing minerals in R-layer than in layers I. 1 and I. 2 (7.4% and 6.8%), in which

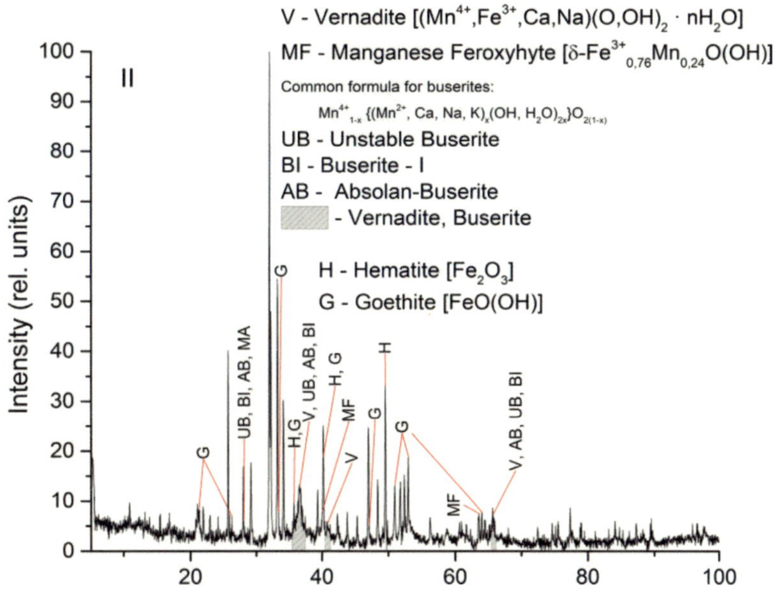

Figure 7.
X-ray diffractograms of layer I-1, I-2.

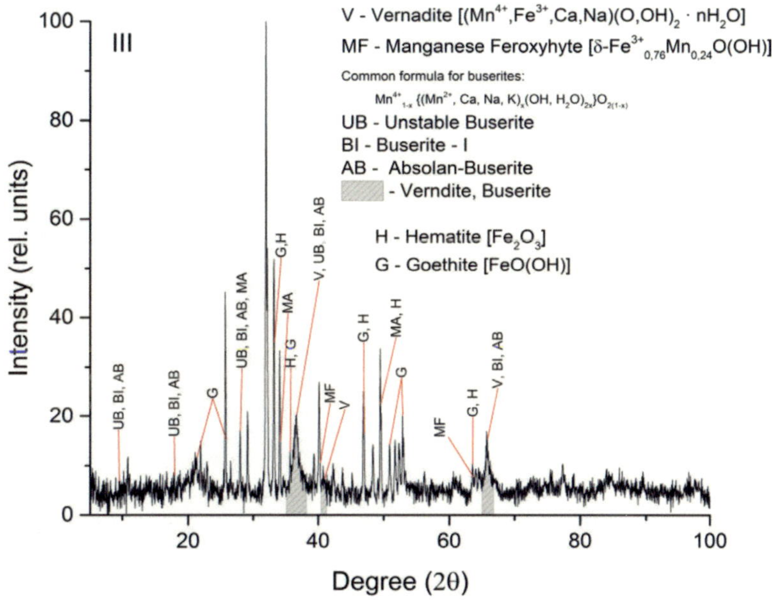

Figure 8.
X-ray diffractograms of layer II, III.

bacterial activity is already observed. The R-layer phase compositions differs from the subsequent layers: in addition to trivalent oxides, it contains amorphous ferrihydrite ($5Fe_2O_3$. $9H_2O$) and divalent (FeO) wustite. In the subsequent layers I. 1 and I. 2, only trivalent iron oxides are present, which is explained by the active activity of iron-oxidizing bacteria. A comparative analysis of the Mossbauer spectra taken at room and nitrogen temperatures allowed us to estimate the size of these superparamagnetic particles. They are lesser than 7 nm for goethite, and lesser than 5 nm for hematite. The obtained estimate of the particle sizes of goethite and hematite corresponds to the sizes of biogenic nanoparticles [12]. Results are shown in **Figure 8**.

Based on the results obtained, it is possible to trace the changes that occurred in the crusts over millions of years and make assumptions about the changes in the external conditions of their formation (**Figures 9** and **10**).

Since the main source of iron in the crusts was the products of volcanic eruption and collapsing basalts, the iron presented in the base of the crust in a metallic or divalent state. After that, under the influence of oxidizing bacteria, iron ions are oxidized to a trivalent state. This explains the presence of low concentrations of divalent wustite in the relict layer and the high content of trivalent iron in the composition of nanosized oxides: amorphous ferrihydrite, goethite, and hematite. In the subsequent layers I-1 and I-2, in which iron-oxidizing bacteria are active and ore-bearing stromatolites appear, an increase in the magnitude of the effect of the goethite and hematite phases is observed, with goethite as the most intensive phase.

In layer II, one can notice significant changes that have occurred with the iron oxides in the crust. Probably, during the global glaciation corresponding to the age of this layer (3.8–2.4 million years), other types of bacteria appeared with changes

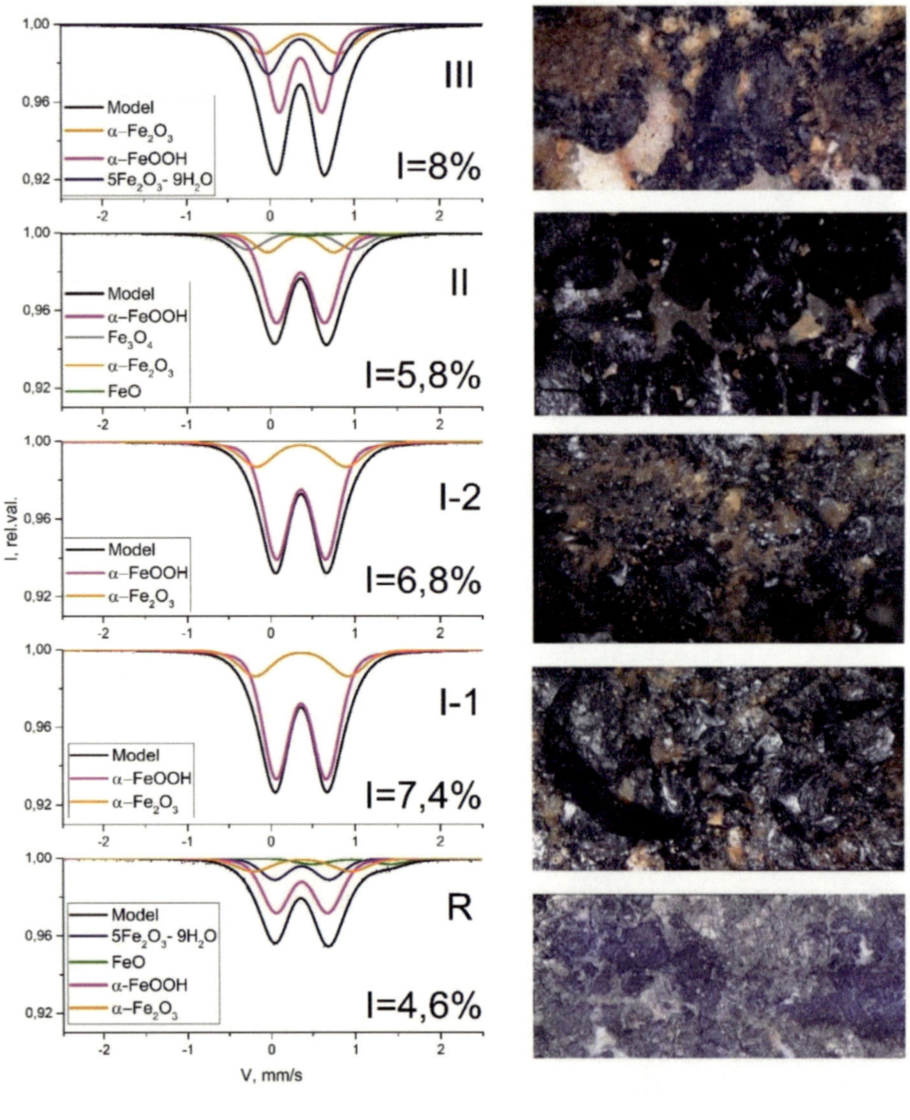

Figure 9.
Mössbauer spectra of various layers of the crust.

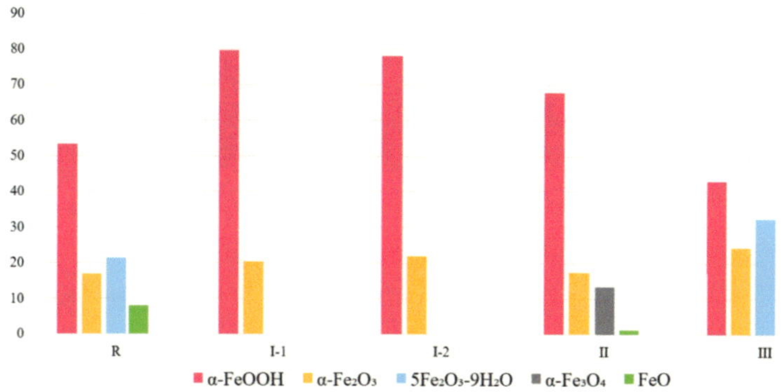

Figure 10.
Quantitative phase composition diagrams of all layers of the studied crust.

in external conditions (changes in pH, temperature, environmental composition, including a large content of decaying endangered organisms). It leads to the trivalent iron contained in nanoscale goethite and hematite transformation to divalent in wustite and partially divalent in magnetite.

In layer III, the phases of goethite and hematite are again observed. Probably, by the time of the formation of this layer, the reduction processes in the environment stopped and the colonies of oxidizing bacteria began to work actively again.

3. Conclusion

A comprehensive layer-by-layer analysis of the ferromanganese crust showed that the crust is practically a composite consisting of clastic and volcanogenic micron-sized minerals and stromatolites, consisting of fossilized biofilms filled with nanoparticles of iron and manganese oxide compounds, which are bacteria waste products. The lower layers of the crust are very dense: they contain many clastic minerals, and ore-bearing stromatolites grow between them. During the growth of the crust, bacteria constantly carry out their vital activity, therefore stromatolites grow continuously, occasionally changing their direction and shape under the influence of external conditions. And minerals: apatite, quartz and others come to the place of crust growth only occasionally - after a volcanic eruption. Therefore, the proportion of the iron-ore and manganese-ore components increases from the lower crust layer to the upper layer, which is almost entirely composed of these components. This leads to the fragility of the upper layers of the crust.

Experimental methods have provided objective evidence of the biogenic nature of the ore components of ferromanganese crusts. And the method of Mössbauer spectroscopy, first applied to these objects, made it possible to obtain a quantitative phase analysis of iron ore components in different layers of the crust. Based on this analysis, it was possible to trace how the composition of iron oxides changes in different layers of the crust, depending on changes in external conditions.

Author details

Alla A. Novakova* and Dmitrii S. Novikov
Department of Physics, Lomonosov Moscow State University, Moscow, Russia

*Address all correspondence to: novakova.alla@gmail.com

References

[1] Hein J.R., Koschinsky A.: Deep-Ocean Ferromanganese Crusts and Nodules. Treatise on Geochemistry 2-nd Edition. 2014; 273-291. DOI 10.1016/B978-0-08-095975-7.01111-6

[2] Banfield J.E., Nealson K.H: Geomicrobiology: Interaction between microbes and minerals. Rev. Mineralogy. 1997; 3: 448. DOI 10.1180/minmag.1998.062.5.01

[3] Lovley D.R.: Bioremediation of organic and metal contaminants with dissimilatory metal reduction. J. Ind. Microfiol.. 1995; 14: 85-93. DOI 10.1007/BF01569889

[4] Avdonin V.V., Zhegallo E.A. Sergeeva N.E. Bacterial nature of oxide ferromanganese ores in the World Ocean. Moscow: GEOS; 2019. 98 p.

[5] Astafieva M.M., Gerasimenko L.M., Zhegallo E.A. et al./ed. Rosanov A.Yu./ Fossil Bacteria and other microorganisms (Atlas of fossil bacteria microphotos). Moscow, Russia: Borrisiak Paleontology Institute; 2011. 172 p.

[6] Sherrell R.M., Field M.P., Ravizza G.: Uptake and fractionation of rare earth elements on Pacific Rise. Geochim. Cosmochim. Acta. 1999; 63: 1709-1722. DOI 10/1016/S0016-7037(99)00182-9

[7] Kim N.V., Novikov D.S., Novakova A.A.: Study of the ore components of biogenic ferromanganese crusts. Scientific letters of Physical department of Moscow State University. 2019; 4: 1940502. (in Russian)

[8] Kuzmann E., Nagy S., Vertes A.: Crytical review of analytical applications of Mossbauer spectroscopy illustrated by mineralogical and geological samples. Pure Appl. Chem. 2003; 75:6:801-858. DOI 10.1351/pac200375060801

[9] Madsen D.E., Cervena-Gontrad L.: Magnetic fluctuations in nanosized goethite grains. J. Phys.: Condens. Matter. 2009; 21:11. DOI 10.1088/0953-8984/21/1/016007

[10] Van der Kraan A.M.: Mossbauer effect studies of surface ion of ultrafine hematite particles. Phys. Status Solidi. 1973; A18:216-226. DOI 10.1002/pssa.2210180120

[11] Cornell R.M., Schwertmann. The Iron Oxides, Structure, Properties. WILEY-VCH; 2003. 661 p. DOI 10.1002/3527602097

[12] Novakova A.A., Dolzhikova A.V., Novikov V.M., Boeva N.M., Zhegallo E.A.: Comparative Analysis of the Structural and Morphological Features of Biogenic and Synthesized Goethite Nanoparticles. Crystallography Reports. 2017; 62:6:971-975. DOI 10.1134/S1063774517060189

Chapter 2

History and Current State of Mining in the Kryvyi Rih Iron Ore Deposit

Mykola Stupnik and Volodymyr Shatokha

Abstract

In 2021 one of the world's largest iron ore deposit in Kryvyi Rih (Ukraine) celebrates 140 years of its exploitation history. During the whole period of its existence the deposit has played and continues to play an important role in the development of Ukraine's economy, being the main basis of its iron and steel industry. More than 6 billion tons of marketable iron ore extracted during this period and some 20 billion tons of waste rock has been mined. The deposit constitutes 82% of Ukraine's iron ore output making the country the 7th biggest producer and 5th biggest iron ore exporter with value of USD 4 billion in 2019. In this chapter the historic aspects of deposit's development and current state of its exploration are analyzed, including processing techniques employed to produce high grade iron ore concentrate, sinter and pellets. Characteristics of iron ores' mineralogical composition and the features of the deposit's geological genesis are also presented. Special attention is paid to the ongoing and planned modernization and deployment of innovative technologies aimed to enhance the competitiveness and to reduce environmental footprint of exploration.

Keywords: iron ore, deposit exploitation, beneficiation, sintering, pelletizing

1. Historic introduction

In 140 years of history, Kryvyi Rih has gone through an impressive path of transformation: from a provincial town in the South of the Russian Empire to the main iron ore base of the Soviet, then – Ukrainian, steel industry, closely involved in the processes of global trade. At present, proven resources of iron ore in Ukraine are estimated at 6.5 billion tons of crude ore (the fifth largest reserve in the world) [1] of which around 70% is deposited in Kryvyi Rih.

The Kryvyi Rih iron ore deposit experienced ups and downs with several distinct periods of formation and development. The settlement of Kryvyi Rih, established as the Cossacks winter-abode in the XVIII century, remained rather unremarkable tiny inhabited area for a long time. The situation changed drastically in the XIX century greatly after the beginning of iron ore industrial mining.

Iron ore deposits were discovered in late 30s of the XIX century. Substantial studies conducted by local entrepreneur Alexander Pol proved the feasibility of the deposit development and in 1880 he initiated the establishing in Paris of the company Societe Anonyme Minerais de Fer de Krivoi Rog with the capital of Franc 30 million owned by French major shareholders (**Figure 1**), which started the iron ore mining in 1881.

Figure 1.
Share of Societe Anonyme Minerais de Fer de Krivoi rog.

Initially, iron ore mining was made almost completely manually (**Figure 2**). Most part of the mined ore was delivered eastwards over the distance of roughly 400 km by horse and cart (including crossing ca 1 km wide Dnipro river by ferry) to the factories of Novorossiysk Coal, Iron & Rail Production Company. This company was established yet in 1869 (first blast furnace was blown-in in 1871) by a British-Russian consortium initiated in the Donetsk region by John Hughes, a Welsh entrepreneur. Transportation conditions remarkably changed after launching of the railway connecting Kryvyi Rih iron ore deposit with coal deposits of Donbass via the bridge across the river of Dnipro commissioned in 1884, which boosted industrialization over the large, previously rural, area along the railroad.

Figure 2.
Iron ore mining in Kryvyi Rih (1899).

Figure 3.
Gdantsevka ironmaking plant (1899).

In 1885 the mentioned above Novorossiysk Company started also iron ore mining in Kryvyi Rih area, based on British funds. It was followed by starting several other mines, owned mostly by international capital represented with Belgium, Great Britain, France and Switzerland. In addition to manufacturing organization, leading foreign experts have been assigned to deliver mining and metallurgical businesses towards the best European practice. Already by 1887, Kryvyi Rih became the dominant iron ore extraction site of the Russian Empire. In 1892 Gdantsevka Ironmaking Plant (**Figure 3**), also owned by Societe Anonyme Minerais de Fer de Krivoi Rog, was put into operation in the outskirts of Kryvyi Rih. The major part of iron ore extracted in Kryvyi Rih deposit was consumed in Russian Empire, and 10-20% was exported.

In the beginning of the XX century as many as 63 mines have been in operation with cumulative depth of more than 100 km [2]. Each mine was a separate unit with its own infrastructure and administration. The largest production volume of 6.2 Mt. was achieved in 1913, constituting 72% of total production of iron ore in the Russian Empire. Almost 20 thousand employees worked at the mines, 40 engineers controlled the production, 7 of them were foreigners.

However, during World War I followed by the Bolshevik Revolution (1917) and civil war lasted through 1921, the extraction of ore practically ceased: during seven months from January to July 1918 just 264 kt of iron ore was supplied, whereas before the war average monthly supply was around 480 kt. By August 1918 all iron ore mines in Kryvyi Rih were stopped and most of them – flooded [3].

2. Post WWI reconstruction and industrialization

After a kaleidoscopic change of ruling powers, Soviet regime was established in Kryvyi Rih region by the beginning of 1920. Iron ore was essential for reconstruction of industry, so local authorities were subordinated directly to Kharkiv, the first capital of the Ukrainian Soviet Republic. Initially, within the Kryvyi Rih deposit the martial law was imposed, but such measure was not successful and therefore it was decided to return to economic management instruments.

Reconstruction took place in very difficult conditions: famine, ruin, lack of materials, and shortage of workers. Initially, only seven mines were brought into operation. In 1927 as much as 3.5 Mt. of iron ore was mined being just 56% from 1913 level.

Since 1927 the new mining technologies have been introduced replacing the widely used cut and fill stoping method. In 1928 the first "Bucyrus" excavators were supplied from the USA along with 12 scraper winches, 7 conveyers and hammer drills of various manufacturers. The iron ore output in 1930 reached 9.78 Mt., 77% of which was produced by underground mining. Fifteen new mines equipped with an up-to-date machinery were built during the First Five-Year Plan period (1928-1932). Further mechanization of mining operations resulted in drastic increase in labor productivity and by 1940 the annual output reached 19 Mt. [4].

The mining was ceased by August 8th, 1941 - few days before Kryvyi Rih was occupied during World War II. To prevent usage of mines by the invaders, most of the equipment was transported to the mining enterprises of Ural. Units weighting over 100 t were blasted. Most of miners were also evacuated. However, more than 16 Mt. of iron ore was left in stocks.

Reconstruction of mining enterprises during German occupation was sluggish, focused mostly on nearby Nikopol manganese ore deposit being essential for the manufacture of advanced steel grades, whereas Kryvyi Rih with its iron ore stock and a huge amount of equipment turned into scrap was used as auxiliary source of iron charge in steelmaking. Demand for iron ore was hindered by slow reconstruction of coal mines in Donetsk deposit and of steelmaking enterprises on the occupied area, destroyed before the Soviet Army retreat. As a result, iron ore mining did not began until the end of 1942 and extraction of high-grade ore was started only in May 1943.

In June 1942, the removal of ore from the stock began, resulting in usage of 347 kt by the end of the year. In addition to this, from April to September 1943 about 385 kt of ore was mined. As the whole, during the period of occupation 1.78 Mt. of iron ore was taken, out of which 1.4 Mt. from the stocks [5].

In November 1943, anticipating Soviet Army's offensive, the invaders destroyed equipment, transport facilities and workshops of almost all reconstructed enterprises. As the result, out of 77 producing and servicing shafts, 54 shaft mouths were exploded or destroyed completely to the depth from 6 to 15 m. So, within 30 months the mines and the equipment have been destroyed twice.

3. Post WWII reconstruction and further development

Reconstruction of industry with the implementation of large-scale projects has started after the liberation of Kryvyi Rih in February 1944. While the miners began to return from evacuation, the government adopted the law stipulating reconstruction of mines and of mining industry's infrastructure. In just in one year, 32 mines were reconstructed and put back into operation. Another 11 large mines and 15 ventilation units were put into operation in 1946 and annual output reached 6 Mt. The plan of reconstruction and development of Kryvyi Rih mines was drawn up aiming to reconstruct and build 35 large and medium-sized mines with a total annual designed capacity of 26 Mt. This plan was successfully fulfilled and by 1950 ore mining exceeded pre-war level.

Initially, to modernize the facilities and improve performance, mostly inexpensive and simple in implementation measures have been applied; however, since the beginning of 1960s, general reconstruction of mines included application of new technologies and equipment aimed at advanced level of mining operations. This included deepening of more than 40 mines to the depth from 200 to 1250 m with the cumulative deepening by about 28 km. Mines with inclined shaft were also built during this period.

Construction of Mining and Processing Plants (MPP), large factories comprehensively encompassing production steps from mining through high grade concentrate (some factories also produced pellets or sinter), manifested beginning of new era of mining development in Kryvyi Rih. Supply of high quality agglomerated iron ore materials also greatly contributed to boost development of steelmaking industry.

The first attempts to beneficiate ferruginous quartzite were made in 1930s at the experimental plant, and studies in this direction have been continued after WWII, so research and industrial expertise had been already in place when the decision to build the Southern MPP was adopted in 1952, which produced its first concentrate in 1956. The success of the enterprise with annual mining and beneficiation capacity of 9 Mt. facilitated further construction of similar enterprises and increase in production capacity of existing plants. From 1959 through 1966 another four MPPs - Novokryvorizkyi, Northern, Central and Ingulets - were built on different sites in and around the city of Kryvyi Rih. All plants had almost identical structure including open pit mines for iron ore extraction and beneficiation plants to produce concentrate (**Figure 4**). All MPPs except of Ingulets have been also equipped with plants producing sinter or pellets. Production of iron ore in Kryvyi Rih peaked in 1978 at 120 Mt., when in the aftermath of economic stagnation the demand for iron ore ceased to grow.

Kryvyi Rih was also one of the major suppliers of iron ore materials to the Central-East European Countries members of former Council of Mutual Economic Assistance (CMEA). In 1984 Soviet Union, East Germany, Czechoslovakia, Romania and Hungary initiated the latest large collaborative project of CMEA - construction of new iron ore beneficiation plant near Kryvyi Rih to produce for the countries involved the iron ore concentrate from oxidized quartzites with annual designed capacity of 9.1 Mt. of concentrate. Romania alone invested US$526 m out of US$1.6bn totally spent. However, following the collapse of communist regimes, the construction has never been finished [6].

Figure 4.
One of the world largest open pits of Ingulets MPP: Depth 520 m, designed capacity by crude ore - 38 Mt./year.

4. Sector development in modern Ukraine

4.1 General development trends

Collapse of the USSR in 1991 broke established supply chains and stopped large infrastructural projects thus substantially reducing demand for steel. Therefore, the iron ore production in Kryvyi Rih in early 1990s plunged down as shown in **Figure 5**. Mining enterprises faced a severe financial situation. Profit and available capital decreased drastically: up to 80% of transactions in the steel sector were made via barter schemes during this period, impeding accumulation of capital [8].

Further drop of iron ore output was prevented in the frames of a large scale "sectorial experiment" covering the entire mining-metallurgical complex of Ukraine and conducted from August 1999 to January 2003, providing enterprises with state assistance and allowing companies to accumulate finances for improving technologies and environmental safety under special taxation regime [8]. The state of Ukraine's mining sector was improved and iron ore output started to rapidly grow, as shown in **Figure 5**. Meanwhile export opportunities have been gradually exploited, especially after 2008, when the domestic demand shrank drastically in the aftermath of financial crisis followed by recession. **Figure 6** compares iron amount of the produced ore with the output of pig iron (blast furnace ironmaking is the sole large consumer of iron ore products - sinter and pellets in Ukraine). The both values, after being fluctuating almost in parallel, since 2009 gradually decouple from each other, and in 2019, first ever, output of pig iron drops by 2.4% although the iron amount of the produced ore grew by 2.8%. This illustrates orientation of mining sector rather not on the domestic needs but on the global markets - as shown in **Figure 7**, since 2015 over 60% of the iron ore output is exported.

4.2 Current characteristics of deposit

In Ukraine, by the exploration stages, reserves are coded differently from International system applied by CRIRSCO (Committee for Mineral Reserves International Reporting Standards) and the **Table 1** for conversion is shown below. Collectively, the sum of A + B + C1 categories is referred to as a proven reserve.

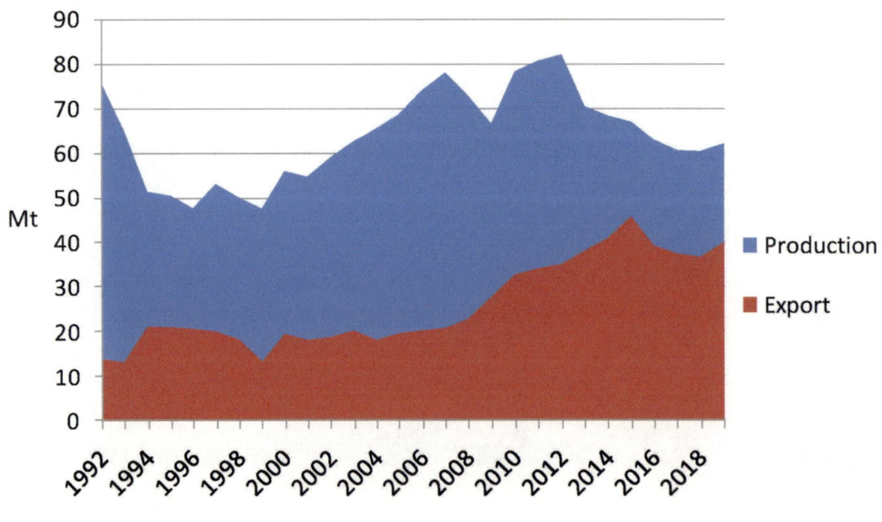

Figure 5.
Production and export of usable ore (elaborated by authors on [7]).

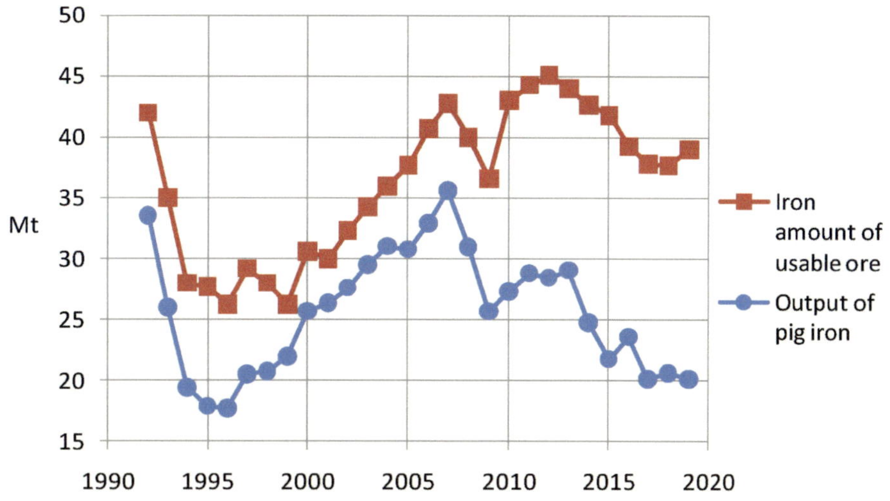

Figure 6.
Iron amount of the produced ore versus output of pig iron in Ukraine (elaborated by authors on [7, 9]).

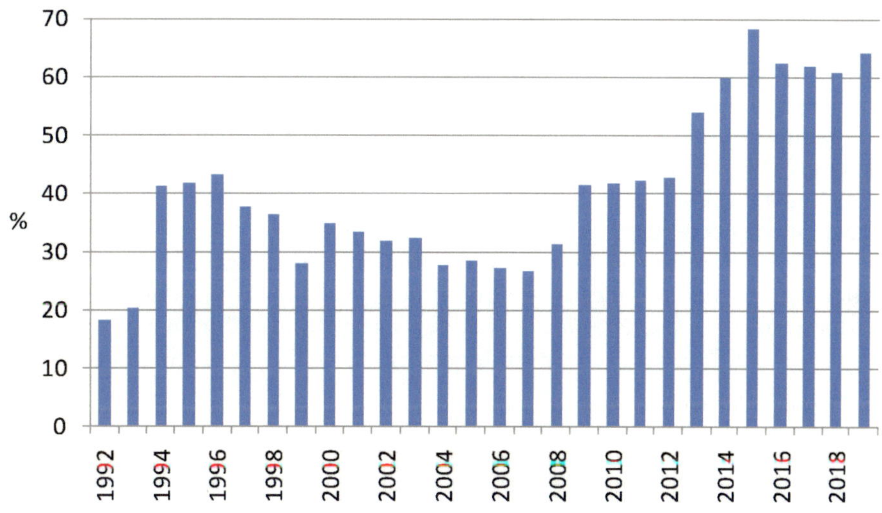

Figure 7.
Share of export in the usable iron ore production.

CRIRSCO	*Classification*	Measured		Indicated	Inferred
	Level of confidence	High		Reasonable	Low
Former Soviet code	*Classification*	A - fully explored	B - studied based on industrial development	C1 - studied based on the results of pilot development and testing	C2 - studied based on the results of testing and exploration
	Level of confidence	Proven			Probable

Table 1.
Table of correspondence between CRIRSCO and former soviet codes.

Over 70% of Ukraine's proven reserve is located in Kryvyi Rih iron ore deposit. Iron ores mainly belong to the following three geo-industrial types: rich magnetite–hematite-martite ores, ferruginous quartzites and brown ironstones. Rich ores are used without concentration. Magnetite and cummingtonite-magnetite quartzites and brown ironstones of Kryvyi Rih deposit are concentrated by relatively simple methods – washing and magnetic separation. Oxidized quartzites and brown ironstones require roasting-magnetic and gravity-floatation methods of concentration.

Rich ores occur in bodies of 10-60 m and, occasionally, up to 100 m thickness. The iron content varies from 46% to 70%, phosphorus and sulfur contents – from 0.01% to 0.03%.

Magnetite and oxidized ferruginous quartzites are encountered in bodies of up to 100-200 m and, occasionally, up to 500 m thickness. The iron content in such ores varies from 14% to 46%, phosphorus content - from 0.03% to 0.16% and sulfur content - from 0.02% to 0.24%.

Currently, approximately 51% of all proven reserve is exploited. Substantial part of the reserve is not considered for mining in the near future for economic reasons. Along with underground enterprises, five large enterprises mine the ore in 10 open pits at depths of 150-450 m with further processing. Since the early 1970s the open pit mining method became and to date remains predominant as shown in **Figure 8**. In terms of maintaining production on current level, available reserves of iron ores covered by ongoing open pit mining activities remain sufficient to ensure functioning of mining and processing plants in the long term perspective as shown in **Table 2**.

Currently, rich iron ores and magnetite quartzites are mined via the underground method by Kryvyi Rih Iron Ore Plant, ArcelorMittal Kryvyi Rih and Sukha Balka. Magnetite and oxidized quartzites are mined using the open pit method by Ingulets MPP, Southern MPP, Northern MPP and Arcelor Mittal Kryvyi Rih, whereas Central MPP applies both underground and open pit mining methods.

The depth of occurrence of commercial reserves of iron ore in Kryvyi Rih reaches 2.7 km, therefore underground mining remains essential. However, mining conditions are rather complicated - eight underground iron ore mines are functioning with mining operations conducted at depths of 600-1300 m in particularly risky underground conditions. Three of underground mines are operating in a flooding prevention mode.

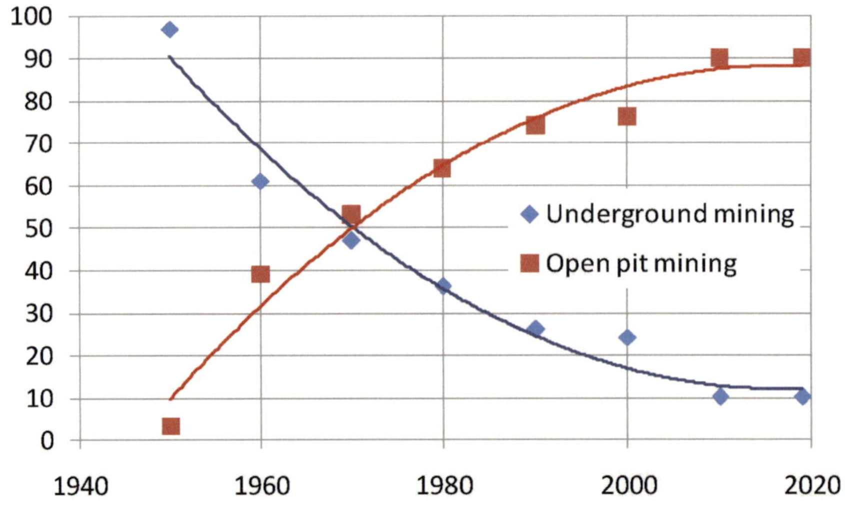

Figure 8.
The ratio between an underground and an open pit iron ore mining in Kryvyi Rih (%).

Reserves	ArcelorMittal Kryvyi Rih		Northern MPP		Ingulets MPP	Central MPP			Southern MPP
	# 2-bis	# 3	Pervomaiskyi	Annivskyi		# 1	# 2	# 3	
Proven reserve	265.619	1126.037	858.345	870.873	1956.064	197.078	209.902	157.202	1045.040
Available within current mining boundaries	236.625	750.084	501.458	372.819	972.060	15.912	214.463	152.472	1045.040
Resource endowment, Years	29	27	28	50	33	39	34	157	35.7
Accessed reserve	29.593	30.503	5.74	1.42	4.42	3.24	2.4	0.9	22.8
Developed reserve	5.795	11.577	3.15	0.51	0.52	1.57	1.5	0.5	0.9
Blocked out reserve	1.556	2.792	0.13	0.16	0.15	0.926	0.77	0.16	0.2

Table 2.
Iron ore reserves covered by ongoing open pit mining activities in Kryvyi Rih, Mt. [10].

Generally, the analysis of the situation of underground mining in Kryvyi Rih reveals a threat of gradual reduction of the resource provision and consequent decline of the commercial potential. Iron ores in the amounts of the billions of tons are accounted at underground mining enterprises as so called balance iron ore reserve (meaning the resource which mining is economically feasible and which parameters satisfy accounting requirements). This resource can provide the basis for mining development for more than 100 years ahead. However, in fact, the situation is far from optimistic for individual enterprises and for the deposit as a whole. Currently, underground mining of naturally rich iron ores is carried out only at seven mines of three mining enterprises and one more underground mine exploits magnetite quartzites.

Overall, during the last 10-15 years, the gap between the designed capacity and actual production of enterprises was reduced due to pressure from the market that forces the enterprises to adapt design solutions of current and new projects in order to minimize expenditures, increase technological efficiency and enhance competitiveness. However the drawbacks are not yet overcome and technology still lags behind the market demands, so capacity utilization of most mines does not exceed 70-80%. Such capacity utilization weakens competitiveness on the global market and undermines Kryvyi Rih deposit's production potential.

In 1990, the share of rich ore in the total iron ore production in Kryvyi Rih was 26.3%, while in 1995 it was reduced to 23.5%, and at the end of 2019 – to 14%. Currently, compared to 2007 (the most successful year in the period from 1995 to 2019), the production volume decreased from 62.98 Mt. to 58.27 Mt. or by 7.5%. A more significant drop is observed in comparison to 1990, making 59%.

Key reasons for production decline of rich iron ores in Kryvyi Rih can be summarized as follows:

1. Continuous and intensive exploitation of deposit as a whole, delineation of individual deposits to greater depths, decommissioning of unprofitable mines, and depletion of reserves result in lower resource endowment of existing mines.

2. Lack of efficient technologies for processing of extracted raw ore result in relatively low iron content of ore (55-57%), high losses (up to 12-15%) and ore dilution (up to 10%) in mining.

3. High rate of mining operations deepening (13-16 m per year) is coupled with limited technical capacity of hoisting installations as well as with general deterioration of mining and geological conditions at deeper horizons are followed by a decrease in the mines' productivity.

4. Financial aspects include constant growth in maintenance costs of existing facilities due to reduced levels of mining, increased mining capital and high development cost at new deeper horizons. Increased cost of fuel, energy and of processing equipment only exacerbate the current situation.

5. Some market-related problems shall be noted such as insufficient competitiveness of marketable products compared with the major global producers, low domestic demand and increasing volumes of sinter grade ore imports.

6. Investments to R&D are continuously insufficient thus impeding delivery of innovative technologies capable for addressing current industrial needs.

The problems listed above are accompanied by the lack of scientifically sound strategy for underground mining on the national level. Due to long term and intensive exploitation of underground mines in Kryvyi Rih, the reserves of naturally rich iron ore significantly decreased and make 309.2 Mt. at the depth above 1500 m. The depth of mining has reached 1200-1400 m and approaches the critical lifting capacity of mines. The average reserve endowment makes 22.4 years, which, at first glance, is sufficient for economically sustainable development of underground mining enterprises. However, taking into account the duration of all the operations required for commissioning of new deeper horizons as well as subsequent involvement of new iron ore material types or transition to a combined two-step mining and hoisting scheme, the development perspective is rather uncertain.

When evaluating the state of the raw material base, we cannot omit the fact that exclusion of low-grade lump ores from the total unprocessed ore supply in order to achieve required quality of marketable products results in even more intense exhaustion of the reserves. The yield of marketable products at different mining enterprises varies in the range from 84% to 86%. But even with this yield, the iron content of marketable iron ore products usually reaches just 57.5-59.0%, which is not sufficient for competitiveness. Noteworthy, in the longer perspective the discharge of substandard raw materials is not only inevitable, but can even increase.

Considering the current state of the deposit, we anticipate that, in the next 5-10 years, the enterprises can face a number of difficulties in maintaining optimal production capacity and, in 15-20 years, a greater reduction in output and closure of some mines. This process is irreversible from technical and economic points of view, and will have sensitive social consequences in the region.

At the same time, as much as 589.7 Mt. of rich iron ore is deposited at depth from 1500 to 2500 m. In addition, 4.2 Gt of easily concentrated magnetite quartzites locates at depth of 30-1500 m (2.2 Gt at operating mines and 2 Gt at temporarily closed mines). Most of magnetite quartzites occur at depth of up to 700 m.

4.3 Environmental aspects

Analysis of the environmental issues of iron ore mining requires separate study and goes beyond the scope of current chapter. Therefore we address this topic here very briefly.

Currently operating mining enterprises of Kryvyi Rih, including those operating in the flooding prevention mode, pump up to 40 Mm^3 of ground water from mines and pits annually, of which highly mineralized water makes 16-17 Mm^3 annually. Nearly 30 Mm^3 of return water is in a continuous technological loop at the 5 MPPs. Tailing ponds (6 operating and 2 closed ones) of the mining enterprises occupy the area of 5 000 ha and contain over 2 billion m^3 of ore concentration wastes. Dumps contain over 13.0 billion t of overburden rocks. Underground voids reach volume of 50 Mm^3. Overall, in Kryvyi Rih iron ore deposit on the area of 300 km^2, a great number of potentially hazardous objects such as underground mines, open pits, tailing ponds, industrial sites, disturbed land and underground voids are located.

4.4 Iron ore quality

The global output of iron ore is in compliance with dynamics of the world economy, yet, in general, development of iron ore mining is quite steady. Total production of crude iron ore has exceeded 3 billion t per year and has tripled since 1999. As anticipated by Fitch in Global Iron Ore Mining Outlook, global crude iron ore production will grow modestly from 3.348 billion t in 2019 to 3.482 billion t in 2028, which represents average annual growth of 0.2% during this period being a significant slowdown from an average growth of 4.5% during 2009-2018 [11]. Therefore, taking into account growing iron ore mining capacities, especially in Brazil, iron ore market shall become even more competitive. In terms of international competiveness, the position of Ukraine amongst the other major global suppliers is rather uncertain. For instance, iron content in sinter grade ore supplied from Brazil is 66-67%, from Australia – 61-62%, from Sweden – 67% with silica content in the range of 0.5-3.5%, whereas iron content in Ukraine's sinter grade ore makes 56-61% with silica content of 18-21.5% [1].

At present, Kryvyi Rih underground mines enhance quality of marketable ore through increasing iron content in the last portion of ore drawn from blocks to 50-52% instead of previously standard 46% as well as through screening out the lean lumps with iron content of 43-48% at the crushing-grading factory. Although this assisted to a general trend of increasing the iron content in the concentrate (evolution of average value for Ukraine is shown in **Figure 9**), application of such methods increases also ore losses in situ and iron losses during concentration and, finally, negatively impacts the enterprise's economy. At the same time, quality of ferruginous quartzites mined at open pits in terms of iron content and content of impurities was not improved. Generally, the quality of marketable iron ore product remains well behind the advanced foreign practices, despite certain improvements achieved at some enterprises [12].

4.5 Insights for the development strategy

In the current situation, the identification of the priority directions for the further development of the Kryvyi Rih iron ore deposit with the involvement of R&D institutions within the framework of the national and regional economic development strategy is on demand. A broad assessment of the actual condition of the raw material base and, first of all, rich iron ore reserves at depth above 1500 m, allows to foresee four possible options for further long-term development of underground mining:

1. Mining of rich iron ores to the depth of the engineering capacities of the existing hoisting facilities (1500-1600 m) with subsequent closure or conservation of the mines.

2. Mining of rich iron ores below 1500 m with the use of two-stage ore body opening-up schemes or by radical reconstruction of hoisting complexes' infrastructure in order to enable the ore hoisting from deep horizons to the surface.

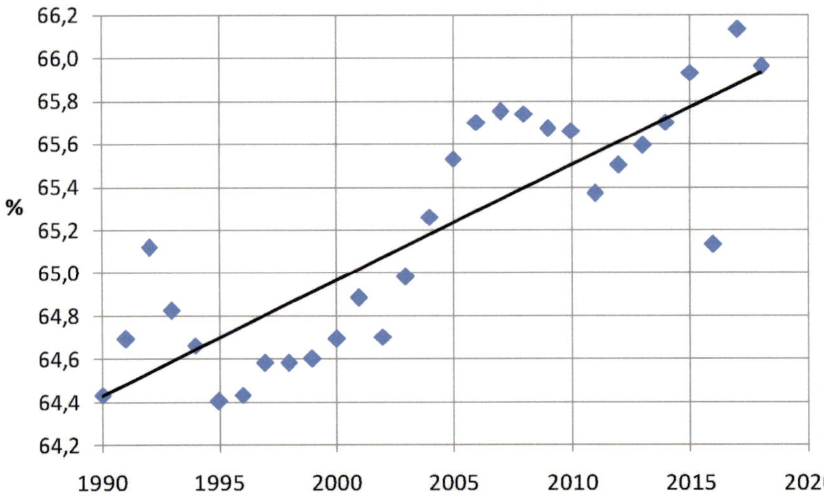

Figure 9.
Average Fe content in iron ore concentrate produced in Ukraine, %.

3. Continuation of rich iron ores' mining above the 1500 m depth followed with the gradual involvement of magnetite quartzites in the areas of operating and temporarily suspended mines and further increasing of the mines' production capacity.

4. Simultaneous mining of rich ores and magnetite quartzites in the fields of operating and temporarily suspended mines.

Since the average supply of rich minerals of underground mining enterprises is only about 22 years, the Option 1 leaves no long-term prospects for the development. For some major mines, for example, Ternovskaya (18 years) and Rodina (17 years), the choice becomes critical. The social consequences of the complete decommissioning of the underground mines throughout the industrial region of Kryvyi Rih will be also very challenging. Moreover, the adjacent territories are likely to be inundated with mining water and water inflows in some pits will also increase. The wet and, in particular, dry conservation of mines requires significant costs for their maintenance. For example, Gigant-Glubokaya and Pervomayskaya mines (both state owned) are currently at the stage of dry conservation and provide flooding prevention for existing mines, pits and adjacent areas. In the case of wet conservation, an irreversible loss of part of the ore reserves, siltation of excavated sites and worsening of the flooding hazard at mining sites and surrounding areas are highly possible.

Options 2, 3 and 4, despite essential difference in the approaches, ensure a certain long term development perspective for mining operations. The advantages are as follows:

1. Retaining of the production and export potential of underground mines and their raw material base is ensured.

2. More holistic approach to development of the iron ore deposits can be applied.

3. Production of high-quality iron ore raw materials through the development of easy-to-process magnetite quartzites present in operating minefields can be increased. The beneficiation of these quartzites using traditional wet magnetic separation technology allows the production of high grade concentrates with iron content of 69-70%.

4. Increasing of production capacity and reducing the hoisting height to the surface may cut the cost of mining in some cases.

However, the choice of the strategy for underground mining shall be based on thorough modeling of the development scenarios for each mining enterprise and for the deposit as the whole. Moreover, the directions of the Kryvyi Rih iron ore deposit development should be seen in the context of both underground and open pit mining, coupled with development of the processing plants. The extraction of magnetite quartzites by the underground method is feasible only in case if the beneficiation facilities sufficient to process the mined ore and to ensure desirable iron content in the concentrate are available. Further deepening of iron ore pits leads to an increase in the haulage distance of rock mass and waste, and, therefore, in the cost of salable products. The gigantic amounts of mined ore and rocks require the additional land allocation for the construction of tailings storage facilities – mostly on highly productive agricultural soils. All of these aspects shall be also taken into account in developing of a strategy for the deposit exploitation.

5. Conclusions

1. Kryvyi Rih iron ore deposit played tremendous role in industrialization of Russian Empire and Soviet Union, and continues to be indispensable part of modern Ukraine's industrial development and the wealth creation.

2. Currently, approximately 51% of all proven reserve is exploited, whereas substantial part of the reserve is not considered for mining in the near future for economic reasons. Since the early 1970s, the open pit mining became and to date remains predominant. Available reserves of iron ores covered by ongoing open pit mining activities are sufficient to ensure functioning of mining and processing plants in the long term perspective.

3. Due to long term and intensive exploitation of underground mines, the reserves of naturally rich ore significantly decreased and make 309.2 Mt. at the depth above 1500 m. The depth of mining has reached 1200-1400 m and approaches the critical lifting capacity. The average reserve endowment of 22.4 years may seem sufficient for further development; however, taking into account the duration of operations required to involve new deeper horizons as well as subsequent involvement of new iron ore types and transition to a combined two-step mining and hoisting scheme, the development perspective is rather uncertain. Scientifically sound strategy for underground mining is lacking.

4. Generally, the quality of marketable iron ore product remains behind the advanced foreign standards, despite certain improvements achieved at some enterprises; therefore, international competiveness of Ukraine amongst the other major global iron ore suppliers is rather uncertain.

5. The directions of the Kryvyi Rih iron ore deposit development should be seen in the context of both underground and open pit mining, coupled with development of the processing plants.

Author details

Mykola Stupnik[1] and Volodymyr Shatokha[2*]

1 Kryvyi Rih National University, Kryvyi Rih, Ukraine

2 National Metallurgical Academy of Ukraine, Dnipro, Ukraine

*Address all correspondence to: shatokha@metal.nmetau.edu.ua

References

[1] U. S. Geological Survey, Mineral Commodity Summaries, January 2020.

[2] Jules Cordeweener. La crise industrielle Russe: Krivoï-Rog, le Donetz, Kertsch. Bruxelles, Paris, 1902, 329 p.

[3] Shatokha V., Tarakanov A. The iron and steel industry in the Ukraine during World War I. Steel+Technology, 1, 2019, pp. 88-32

[4] Kryvyi Rih iron ore deposit. To 125-anniversary from the start of industrial iron ore mining. Vilkul Y., Doyar L., Dioadechkin M et al. Publishing center of Kryvyi Rih Technical University, 2006. 584 p. (in Russian language)

[5] Vilkul Y., Ischenko N., Rukavitsyn I., Stupnyk M. Iron ore mining in Kryvyi Rih – sight 135 year's long. History in photographs and documents 1881 – 2016. Knyha plus, 2016. - 124 p (in Russian language).

[6] Shatokha V. Post-Soviet issues and sustainability of iron and steel industry in Eastern Europe. Mineral Processing and Extractive Metallurgy (Trans. Inst. Min. Metall. C). 2017, 126 (1-2) pp. 62-69 http://dx.doi.org/10.1080/03719553.2016.1251750

[7] National Minerals Information Center. https://www.usgs.gov/centers/nmic/iron-ore-statistics-and-information Accessed: October 28, 2020.

[8] Hanzl, D. and Havlik, P. 2004. Metals sector in CEECs and Russia: a comparative analysis in the European context. East-West Journal of Economics and Business, 7, pp. 81-99.

[9] World Steel Association. Statistical reports. https://www.worldsteel.org/steel-by-topic/statistics/steel-statistical-yearbook.html Accessed: October 28, 2020

[10] Vilkul Yu. G., Azaryan A.A., Azaryan V.A., Kolosov V.A. Quality of iron ore raw materials of underground and surface mining as a basis for mining industry of Ukraine competitiveness. *Metallurgical and ore mining industry*. 2012, 5, pp. 1-4 (in Russian language).

[11] Global Iron Ore Mining Outlook. Fitch Solutions (17 May, 2019) https://www.fitchsolutions.com/corporates/metals-mining/global-iron-ore-mining-outlook-17-05-2019 Accessed: October 28, 2020

[12] Kolosov V.A., Golyarchuk N.I. Problems of enhancement of iron ore raw material quality in Ukraine. *Quality of mineral raw materials: Collection of scientific papers.* Kryvyi Rih. 2005. pp. 36-45 (in Russian language).

Section 2

Characterisation and Processing

Chapter 3

Magnetic Separation of Impurities from Hydrometallurgy Solutions and Waste Water Using Magnetic Iron Ore Seeding

Haisheng Han, Wenjuan Sun, Wei Sun and Yuehua Hu

Abstract

The removal of iron ion from leaching solution is critical for the recovery of value metals, with the method of choice commonly being crystallization (precipitation). This paper summarized the new improvements in iron removal by precipitation methods in recent years and proposed a novel process, magnetic seeding and separation. The new process can promote iron precipitate aggregation and growth on the surface of the magnetic iron ore seeds. A core-shell structure was formed of iron precipitate and magnetic iron ore seeds, which can be magnetized and coalesced in magnetic field, accelerating the solid-liquid separation. The efficient magnetic flocculation and separation offset the poor settleability and filterability of the residues, contributing to the development of the hydrometallurgy process. Moreover, magnetic seeding and separation was also used for the removal of organic and inorganic contaminants from wastewater, significantly improving the purification efficiency. Therefore, iron ore not only played an important role in mining and steel manufacture, but also can be used to solve some problems in crossing fields.

Keywords: leaching solution, goethite process, iron removal, magnetic seed

1. Introduction

Iron is one of the most abundant elements in the earth's crust. It always coexists with metals in the ore, mainly exists in the form of hematite, magnetite and muscovite on the surface of particles or in the inclusions inside crystals [1]. In hydrometallurgy, iron, although is converted into insoluble precipitates and removed in advance by sulfation roasting, soda roasting, acid leaching, etc. during ore pretreatment, still inevitably goes to the aqueous solution with the dissolution of the target metal during the leaching process [2–4]. The classical methods for removing iron in the leaching solution are precipitation, extraction, ion exchange, displacement, and electrowinning [4]. The commonly used method is the precipitation method, which separates iron ions by converting to iron precipitation compounds. According to the different iron precipitation compounds, it can be divided into jarosite [5–6], hematite [7], iron(III) oxide-hydroxide [8] and goethite [9–10] method, etc. The jarosite method produces a large amount of low-grade iron-bearing slag in the application, which is difficult to handle, consumes a large amount of sulfate, and causes certain environmental problems [5–6]; the hematite method needs to be carried out under high temperature and pressure, which consumes large energy and high CAPEX (capital expenditure) [7]. The filtration efficiency of $Fe(OH)_3$ colloid precipitation method is low, and it is easy to adsorb a large amount of other valuable metals, causing large metal loss [8].

The goethite method is widely used in hydrometallurgical plants for zinc, copper and nickel as the main process for removing iron because of its low CAPEX and environmentally friendly products [9–10]. In order to ensure the effect and efficiency of iron removal, the goethite process must strictly control the concentration of Fe^{3+} below 1 g/L, and thus developed the two commonly used processes - VM method and EZ method [8–9, 11]. The former firstly reduces all the iron ions to Fe^{2+}, and then slowly oxidizes the Fe^{2+} to Fe^{3+} under hydrolysis conditions to control the content of Fe^{3+} [9], and the latter slowly adds the concentrated pressure leachate containing Fe^{3+} in the precipitation vessel with addition rate of less than the Fe^{3+} hydrolysis rate, thereby forming goethite precipitation [11]. The pH in goethite process is common lower than 4.0, and calcium hydroxide or calcium carbonate is usually used as neutralizer, which will result in a large amount of calcium sulfate mixed with the goethite residue [12]. These mixed residues reduce the filtration efficiency and cause the loss of valuable metals such as Zn and Ni [5, 13–14]. In addition, the residue mixture accumulated in the tailings pond contains heavy metals such as Pb, As, and Cr, which causes pollution of local water and soil. Therefore, improving filtration performance and reducing the loss of valuable metals are two problems that need to be solved urgently in the traditional goethite precipitation method.

This article summarizes the new improvements in iron removal by precipitation methods in recent years, and on this basis, proposes a novel iron removal process - magnetic seeding and separation. A core-shell structure is formed by precipitating and growing iron on the magnetic seeds surface, and achieves high-efficiency solid-liquid separation by magnetic separation. The new process remarkably reduces the loss of valuable metals in iron removal. Magnetic seeding and separation processes have not only been successfully used in the removal of iron from hydrometallurgical leachate, but also shown good application prospects in wastewater and soil pollution treatment.

2. Iron removal in nickel and zinc leaching solution

2.1 Magnetite precipitate process

Magnetic flocculation and separation based on the magnetic difference of materials can easily separate magnetic solids from mixtures. It shows higher selectivity and efficiency than centrifugation and filtration, and has been widely used in water treatment, biotechnology and minerals separation [15–18]. As is shown in **Figure 1**, Han et al. [3] studied the feasibility of magnetite precipitation in the hydrometallurgical nickel leaching solution. Under lower oxidation potential, at pH 2.0–2.2 and 90–100°C, the iron ions in the leachate may slowly oxidize and partially precipitate in the form of magnetite. Magnetic flocculation and separation can effectively separate the precipitate from the solution. Unfortunately, the iron precipitation from solution is still dominated by goethite, the magnetite composition is relatively small, and it is difficult to truly achieve effective magnetic separation in industrial applications. But this research of magnetite process of iron removal provides other new ideas of magnetic separation.

2.2 Induced crystallization goethite process

The goethite process can be divided into four stages: (a) hydrolysis to monomers and dimers; (b) the reversible stage involving rapid growth to small polymers; (c) formation of slowly reacting large polymers; and (d) precipitation of a solid phase [19–20]. The goethite precipitation system is a complex system, and the presence and content of different components and iron phases have a greater impact on the precipitation and filtration performance of goethite. As shown in **Figure 2**, the pH and temperature conditions of the sulfate-containing solution determine the existence and content of different iron phases such as hematite, goethite, iron hydroxide and hydroxyl salt [21]. The goethite residues that cause

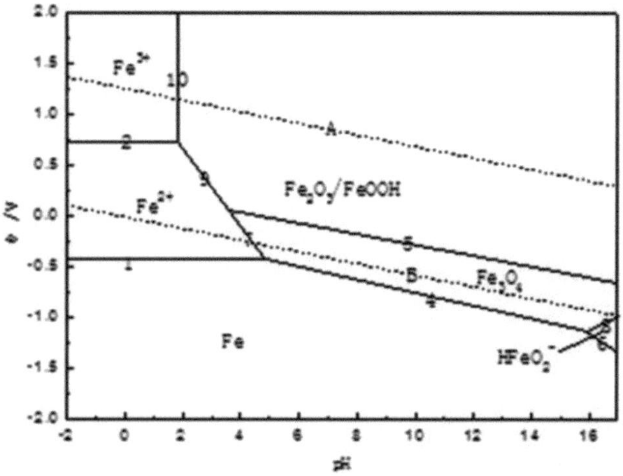

Figure 1.
Potential/pH diagram for Fe-H_2O system at 100°C.

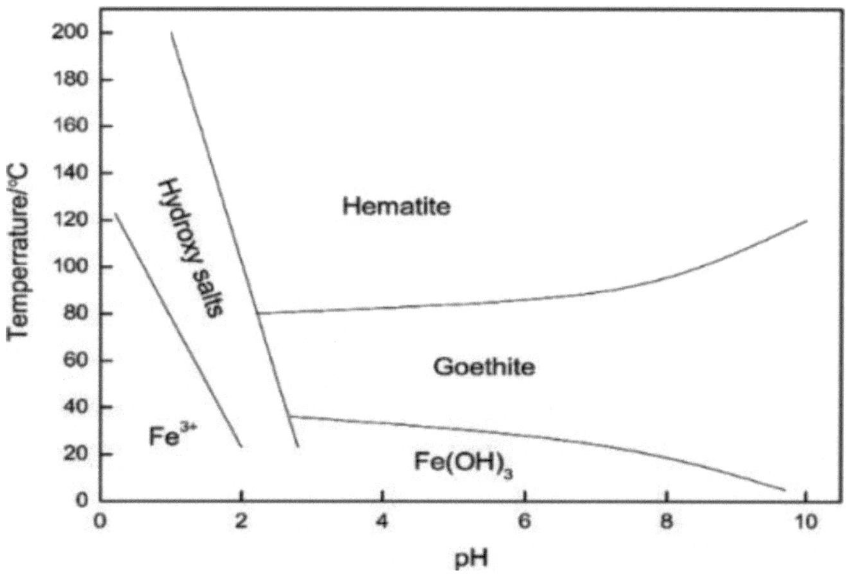

Figure 2.
Temperature and pH conditions for the precipitation of hematite, goethite, ferric hydroxide, and hydroxy salts (including jarosites) from 0.5 M ferric sulfate solution [21, 23].

filtration difficulties and metal loss are composed of amorphous iron phase, six-line ferrihydrite, poor crystalline goethite, solid solution jarosite phase and silica [2, 22]. Therefore, the crystallinity, size and content of the goethite particles can be controlled by adjusting the pH, thereby improving the separation performance and the loss of valuable metals.

Yue and Han [23] study that as the pH value decreases from 5.0 to 2.0, as shown in **Figure 3**, the crystallinity of goethite decreases, the goethite particles tend to agglomerate, the particle size increases significantly, and the filterability of the precipitate improves. Nickel is lost in the iron precipitate by being incorporated into the crystal lattice and adsorbed on the surface of the goethite particles, and the nickel adsorption loss are related to the specific surface area of the goethite particles. When goethite is in an intermediate transition state at low pH (2.5–3.3), which is between the crystalline state and the colloidal state, the loss of nickel is the least. However, the improvement by only adjusting the pH of the goethite precipitation process is minimal. Chang et al. [24] carefully reduced the pH from 4.0 to 2.5, and the loss of nickel is only reduced by about 10% in the iron precipitation. Moreover, it is not realistic to achieve such detailed condition control in actual industrial applications.

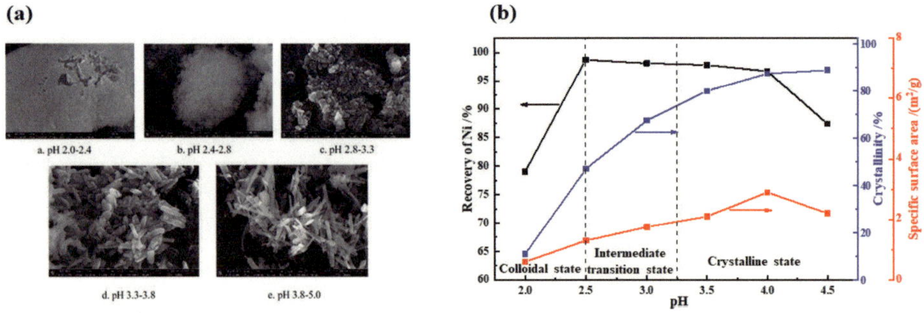

Figure 3.
(a) SEM images of the goethite precipitate at different pHs and (b) pH effect for the nickel loss, the crystallinity, and the specific surface area of the precipitate [23, 25].

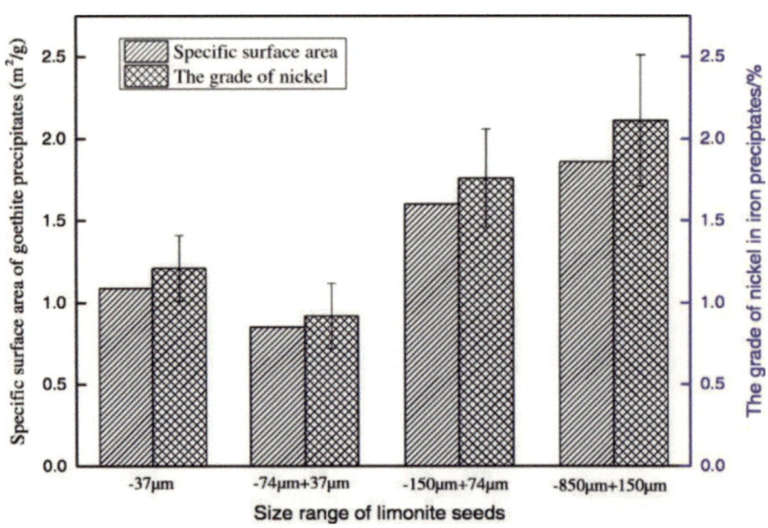

Figure 4.
The specific surface area and the nickel grade of iron precipitates with limonite seeds in different size ranges (2 g/L limonite seeds, pH 2.1–2.5, 85°C) [30].

The traditional goethite precipitation method needs to overcome high barriers to the formation of crystals, and often requires a few days of reaction time. The amorphous iron phase appears at this stage, making precipitation separation difficult. Seed induced crystallization can make crystals precipitate and crystallize from the solution at lower solution saturation, pH value and temperature, and has been widely used in the preparation and production of drugs and nanomaterials [26–29]. Han [30] choose natural limonite as the seed crystal of goethite and induce crystallization to improve the problem of poor filterability at the low pH goethite precipitate. As is shown in **Figure 4**, by adding limonite seeds, the particle size of the goethite precipitate is significantly increased. The goethite particles in the particle size range of 37–74 μm have the largest yield and the smallest specific surface area, which can result iron precipitates with a nickel grade of <1%. However, the reduction of metal loss and improvement of filterability are difficult to achieve at the same time by pH control and induced crystallization, one of them must be sacrificed. The intermediate transition state goethite with good filtration performance and minimum metal loss is difficult to accurately induce formation in the actual field industry. It is a need to find other ideas to achieve qualitative progress.

3. Magnetic iron seeding and separation

Han et al. [25, 31–32] combined seed induced crystallization and magnetic separation, and proposed a novel magnetic seeding and separation process, as shown in **Figure 5**. Before the iron is precipitated as goethite, fine-grained maghemite or magnetite particles are added to the leaching solution to make the goethite precipitate and grow on the surface of the magnetic particles, thereby avoiding mixing with the calcium sulfate precipitation in the solution. The iron precipitates on the surface of the goethite to form large magnetic particles with a core-shell structure, and the precipitates are efficiently settled and separated by magnetic separation. The results show that the iron content in the dry iron residue is more than 52% and the Ni content is less than 0.6%, which can be used in industrial applications to deal with a large amount of iron precipitation. After the calcium sulfate precipitation is roasted, 99% of S and As can be removed, and the roasting residue can be respectively used as raw materials for ironmaking and building materials.

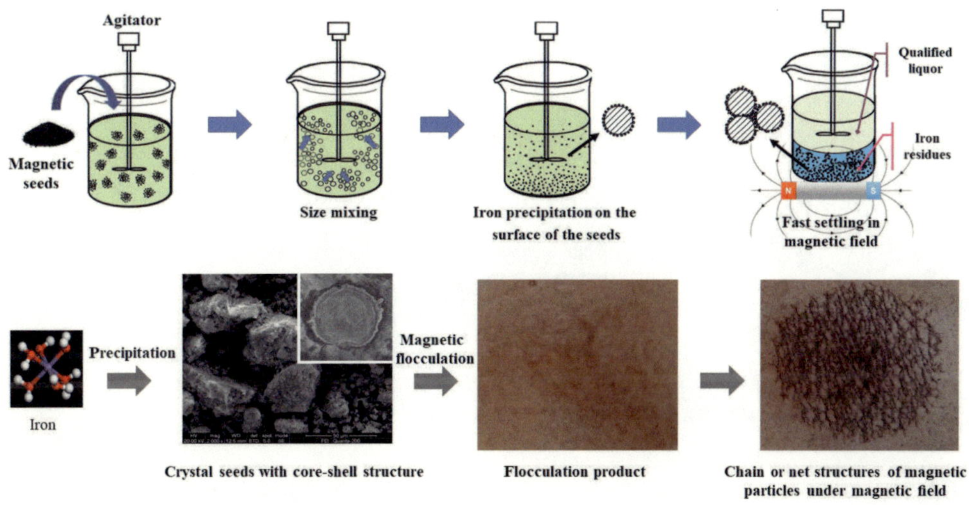

Figure 5.
The process of iron precipitation on the magnetic seeds and the magnetic flocculation in magnetic field [3, 25].

Yue et al. [31] applied magnetic iron seeding and separation to separate goethite from calcium sulfate in zinc leaching with maghemite fine particles as carrier. As is shown in **Figure 6**, the magnetic goethite-maghemite aggregates were separated effectively from calcium sulfate precipitates by magnetic drum separator, and 90% of Fe and Ca is respectively recovered in two corresponding products. Roasting goethite precipitate with coal powder under the optimum conditions removed 99% of S and As. Goethite products can be directly used in the ironmaking industry, and calcium sulfate precipitation can also be used to produce cement and building materials.

Yue et al. [32] establish the surface complex and precipitation model of goethite on magnetite and maghemite magnetic nanoparticles, as shown in **Figure 7**. The formation of Fe (III) surface complexes are directly related to the nucleation and

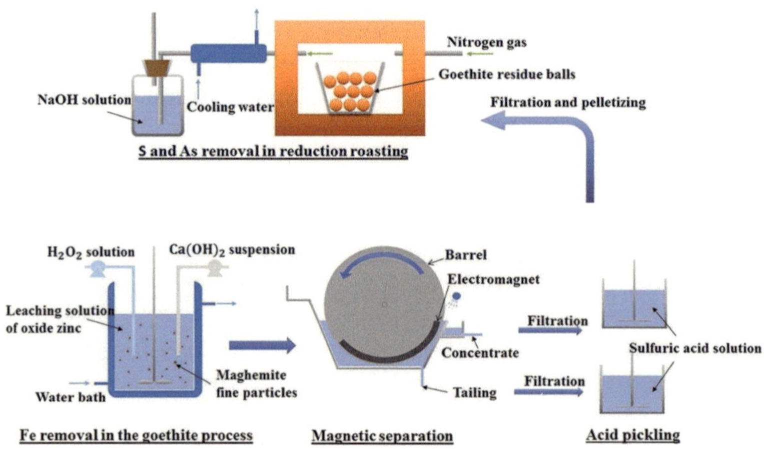

Figure 6.
Schematic illustration of magnetic separation and production of desired goethite and gypsum product [31].

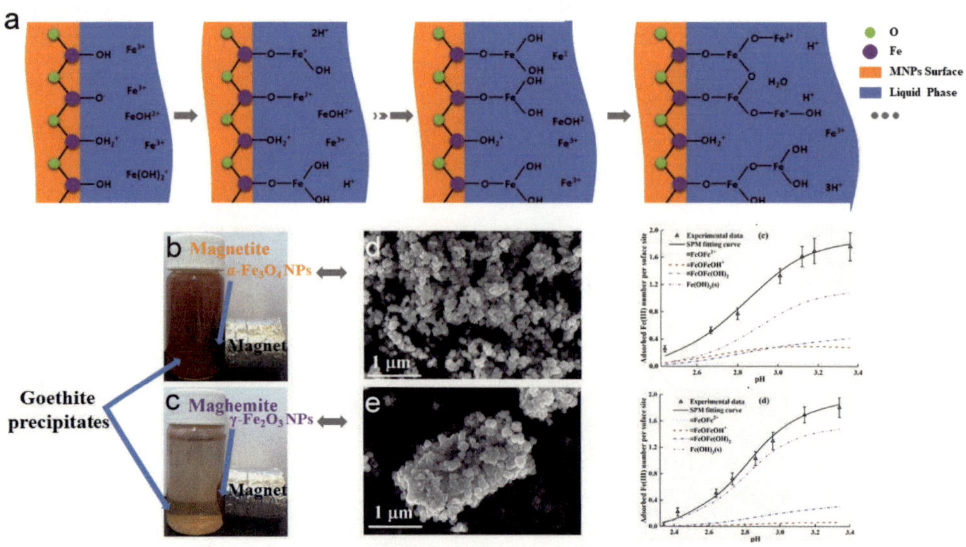

Figure 7.
Surface precipitation model modeling (a) of Fe^{3+} adsorption/precipitation on magnetite and maghemite with corresponding magnetic separation of goethite, images of the suspensions in a magnetic field with 2 g/L (b) magnetite and (c) maghemite NPs, and SEM images of goethite precipitates with (d) magnetite and (e) maghemite NPs [32].

precipitation of goethite on the solid surfaces of the two magnetic nanoparticles. The more polynuclear surface complexes produced on the particle surface, the more precipitation of heterogeneous forms. Fundamentally, it is possible to screen out the best material as the crystal nucleus to separate goethite from calcium sulfate or other heterogeneous precipitation.

4. Application and prospect

4.1 Recycling Fe and Cr in Cr-bearing electroplating sludge

The Cr-bearing electroplating sludge is produced from the treatment of Cr wastewater and metallurgical processes [33–36]. It contains excessive amounts of heavy metals, such as Cr, Fe, Ni, Cu, Pb and Zn, or potential dioxin pollutants [37–38], therefore must be treated before stacking. Many methods have been applied to recover Cr from the acid leaching solution of electroplating sludge, such as electrochemical precipitation (ECP) [39], selective extraction [35, 40], adsorption or biosorption [41–44] and Cr-Fe coprecipitation [45–48]. Compared with other methods, recovering Cr by Cr-Fe coprecipitation is simple, economical and practical for industrial applications. In addition, the advance coprecipitation of Fe and Cr can avoid their interference on the recovery of Ni, Cu and Zn.

Yue et al. [49] use the novel magnetic seeding and separation process to recover Cr(III) and Fe(II) synchronously by forming the Cr(III)-Fe(III) coprecipitates on the surface of maghemite (γ-Fe_2O_3) fine particles. The active hydroxide radicals on the surface of magnetic seeds induce the nucleation and growth of goethite, which results in enhanced Cr (III)-Fe(III) coprecipitation. As shown in **Figure 8**, the maghemite particles, served as the crystal nuclei, could induce the formation of the core-shell structured Cr (III)-Fe(III) coprecipitates on its surface and accelerate the sedimentation of the coprecipitates in the magnetic field. The results of the two-stage coprecipitation showed that the total recoveries of Cr and Fe were 96.17 and 99.39%, respectively, and the grades of Ni, Cu, and Zn in the precipitates were 0.41, 0.38, and 0.22%, respectively. The obtained coprecipitates can be recycled as the feed material of chromium smelting after heat treatment. This method is simple and efficient for high-concentration Cr^{3+} solution treatment, which is beneficial for the sustainable development of resources and environment.

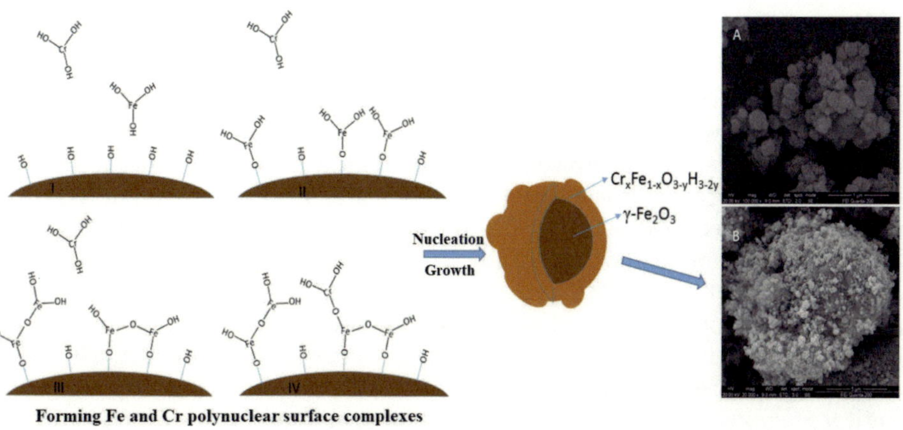

Figure 8.
SEM images of the Cr(III)-Fe(III) coprecipitates without maghemite fine particles (a) and with maghemite fine particles (b), respectively; scheme (c) of the formation of γ-$Fe_2O_3/Cr_x\ Fe_{1-x}OOH$ *with core-shell structure [49].*

4.2 Removal of As in arsenic alkali residue

Arsenic (As) is contained in most metal deposits, and therefore a large amount of arsenic-containing wastewater, flue gas and residues will be produced in mineral processing and smelting, posing a huge threat to the environment [50–52]. Commonly used methods for removing arsenic from solution include precipitation, electrocoagulation, ion exchange, membrane technology and adsorption [53–56]. In order to remove arsenic and recover valuable metals at the same time, these methods all require acid leaching of the waste, which will produce highly toxic and deadly arsine gas [54, 57]. As is shown in **Figure 9(a)**, Yue [58] developed a safer alkaline leaching method - oxidation alkali leaching of the wastes to transform arsenic compounds into arsenate (AsO_4^{3-}) and subsequently recycling the alkali solution after arsenate removal, to treat the arsenic bearing wastes at a lower risk level.

There are a large number of reports that iron oxides have excellent adsorption and precipitation effects on heavy metal ions impurities in aqueous solutions, such as CrU and As. García-Sanchez et al. [59–60] found that goethite has a special adsorption effect and capacity for As ions. Wei Jiang [61] considers that arsenic [AsO_4^{3-}] absorbs on the surface of goethite by forming a bidentate-binuclear complex, and that pH and other metal ions in the solution will affect the distance and coordination number of As/Fe. His et al. [62] found that Uranyl can be adsorbed on goethite, amorphous ferric oxyhydroxide, and hematite sols at 25°C, and the adsorption effect on amorphous iron oxide is the strongest. Yue et al. [58] synthesized a series of high-concentrated ferric oxyhydroxide gels (HFGs) at different supersaturation to adsorb arsenate at high alkalinity, achieving zero-consume of the alkali resources. As is shown in **Figure 9(b)**, using HFG(I) that synthesized under the lowest super-saturation condition as the sorbent to treat the oxidation alkali leaching solution of the copper slag from real industry, the residual concentration of arsenic (As (V)) could decrease from 2084 to 71.8 mg/L, which fully meet the requirements for high-concentrated arsenic stabilization at high alkalinity and alkali resource recycling. To further improve the efficiency of filtration and separation, magnetic seed sowing and separation technology can also be introduced to make this process more complete. Related research is underway.

4.3 Removal of phosphate and starch in wastewater

Phosphorus and starch reportedly are the main wastewater contaminants that are difficult to remove efficiently [63–64]. When the phosphorus concentration in water exceeds 0.02 mg/L, phosphorus becomes a polluting element and causes eutrophication of water bodies [65–67]. Starch is a commonly used and cheap material, widely used in many chemical and material industries, but it produces high concentration of organic wastewater, which will affect the environment [68–70]. Therefore, phosphate and starch removal from wastewater has become the focus of many studies. The main phosphate and starch removal methods are similar, such as chemical precipitation [71–74], biological methods [75–79] and adsorption techniques [80–83]. Among them, Chemical precipitation and adsorption technology is commonly used in wastewater treatment due to the simple operation with low cost and large processing capacity compared to other methods [84–86]. However, chemical precipitation inevitably produces a large amount of fine precipitation and suspended solids, which seriously affect the sedimentation and filtration efficiency [84, 87]. And the adsorbents currently used in adsorption technology, such as activated carbon [70, 88], silica gel [89–90], membranes [91–93], etc., have high production costs and poor adsorption performance, which greatly limits the adsorption effect and industrial applications.

Magnetic flocculation is an effective way to remove ultrafine suspended solids in water treatment [94–95]. It adds magnetic seeds to the aqueous solution to form magnetic flocs with the ultrafine suspended solids in the wastewater, and then passes through a magnetic separator to achieve rapid precipitation and separation [3, 95–96]. The combination of magnetic flocculation and chemical precipitation can make up for the shortcomings of ultrafine suspended solids and low separation efficiency of chemical precipitation. Magnetic flocculation has been widely used to treat wastewater with high pollution concentration [71], high turbidity [96] and high chemical oxygen demand (COD) [97]. It is worth noting that in many studies, iron-bearing minerals have shown the characteristics of removing phosphorus from aqueous solutions [98–99]. The iron-bearing minerals can be coordinated with phosphate and therefore have the potential to be used as adsorption materials for phosphorus and starch in wastewater [100–101].

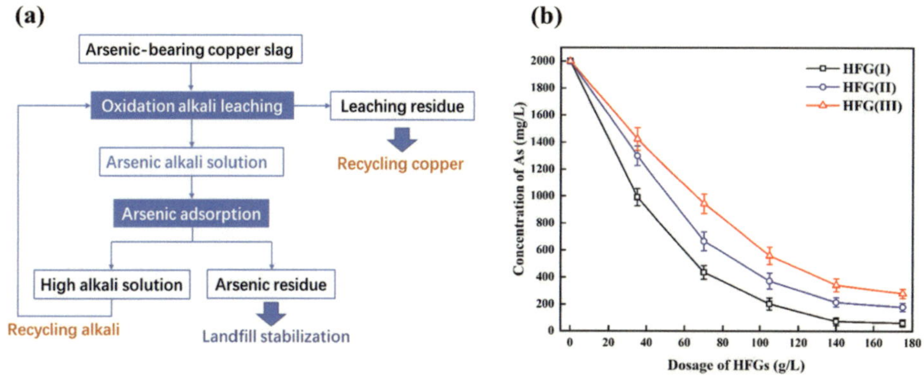

Figure 9.
(a) Flow diagram of the comprehensive treatment of the arsenic alkali residue and (b) arsenic removal from arsenic alkali solution with different HFG samples synthesized at pH 3(I), 7(II), and 11(III) [58].

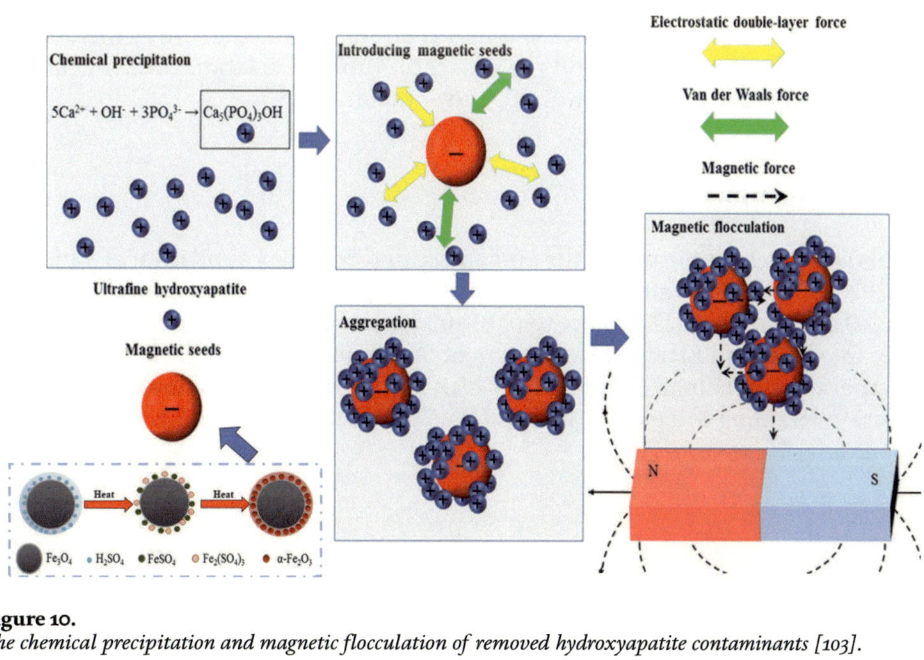

Figure 10.
The chemical precipitation and magnetic flocculation of removed hydroxyapatite contaminants [103].

Du et al. [102–103] combined the magnetic flocculation technology with iron-containing materials to prepare porous magnetic seeds with core-shell structure, which achieved simultaneous removal of starch and phosphate in wastewater. As shown in **Figure 10**, the core-shell magnetic seeds prepared by sulfation roasting of fine magnetite particles have a porous α-Fe_2O_3 structure on the surface, and the specific surface area is three times larger [103–106]. As shown in **Figure 10**, the phosphate and starch in the wastewater can be adsorbed on magnetic seeds surface, and then separated from the wastewater by magnetic separation. The phosphorus and starch content in the wastewater are reduced to 1.51 and 9.51 mg/L, respectively, and the removal rate reaches more than 75% [102].

5. Conclusion

The iron removal method of the hydrometallurgical leachate is still dominated by the goethite process. The goethite process faces the disadvantages of high loss rate of valuable metals and difficulty in separation and filtration, which must be solved to get qualitative improvement. Careful adjustment of the pH value can help reduce metal loss, and inducing crystallization can increase the crystallinity of goethite and improve the separation and filtration efficiency. However, both methods can only focus on solving one of the problems and cannot reduce loss and promote filtration at the same time. The magnetite produced during the precipitation (crystallization) process opened a new path for magnetic separation, while the magnetite method is currently limited to laboratory research. In the present paper, the authors combined the goethite precipitation (crystallization) method with magnetic seed separation technology and developed a novel route. Goethite precipitates on the surface of the external magnetic seeds to form core-shell structured particles, which are efficiently separated by magnetic separation, and at the same time solve the two major problems of the traditional goethite process. This new method also shows advantages in the fields of arsenic and chromium removal from the leachate, phosphorus, and starch removal from wastewater and other fields. Goethite is the most common and stable crystalline iron oxide in soil and sediment. We expect that the goethite method combined with magnetic seed separation technology will show better results in the removal of organic dyeing, heavy metal ions, anions in wastewater and soil, and the adsorption and passivation of chemicals, nutrients, and harmful compounds in environments.

Acknowledgements

This work was supported by the Hunan Natural Science Foundation of China (No. 2020JJ5727), Innovation Driven Plan of Central South University (No. 2018CX036), National 111 Project (No. B14034), and Collaborative Innovation Center for Clean and Efficient Utilization of Strategic Metal Mineral Resources, Key Laboratory of Hunan Province for Clean and Efficient Utilization of Strategic Calcium-containing Mineral Resources (No. 2018TP1002).

Author details

Haisheng Han[1,2*], Wenjuan Sun[1,2], Wei Sun[1,2] and Yuehua Hu[1,2]

1 School of Minerals Processing and Bioengineering, Central South University, Changsha, China

2 Key Laboratory of Hunan Province for Clean and Efficient Utilization of Strategic Calcium-containing Mineral Resources, Central South University, Changsha, China

*Address all correspondence to: hanhai5086@csu.edu.cn

References

[1] Wu Y, Tan D, Ding W, Guo S. Goethite process for removing of iron in hydrometallurgy. Hydrometallurgy. 2014;**000**(002):86-89

[2] Loan M, Newman OMG, Copper RMG, et al. Defining the paragoethite process for iron removal in zinc hydrometallurgy. Hydrometallurgy. 2006;**81**(2):104-129. DOI: 10.1016/j.hydromet.2005.1.1.002

[3] Han H, Sun W, Hu Y, et al. Magnetite precipitation for iron removal from nickel-rich solutions in hydrometallurgy process. Hydrometallurgy. 2016;**165**: 318-322. DOI: 10.1016/j.hydromet.2016.0.1.006

[4] McDonald RG, Whittington BI. Atmospheric acid leaching of nickel laterites review. Part I. Sulphuric acid technologies. Hydrometallurgy. 2008;**91**(1-4):35-55. DOI: 10.1016/j.hydromet.2007.11.009

[5] Claassen JO, Meyer EHO, Rennie J, et al. Iron precipitation from zinc-rich solutions: Defining the Zincor process. Hydrometallurgy. 2002;**67**(1-3):87-108. DOI: 10.1016/S0304-386X(02)00141-X

[6] Swarnkar SR, Gupta BL, Dhana Sekharan R. Iron control in zinc plant residue leach solution. Hydrometallurgy. 1996;**42**(1):21-26. DOI: 10.1016/0304-386X(95)00077-T

[7] Ismael MRC, Carvallo JMR. Iron recovery from sulphate leach liquors in zinc hydrometallurgy. Minerals Engineering. 2003;**16**(1):31-39. DOI: 10.1016/S0892-6875(02)00310-2

[8] Dutrizac J. An overview of iron precipitation in hydrometallurgy. In: Strathdee GL, Klein MO, Melis LA, editors. Crystallization and Precipitation. 2nd ed. Pergamon Press; 1987. p. 259-283. DOI: 10.1016/B978-0-08-035751-5.50038-6

[9] Davey PT, Scott TR. Removal of iron from leach liquors by the "goethite" process. Hydrometallurgy. 1976;**2**(1):25-33. DOI: 10.1016/0304-386X(76)90011-6

[10] Pradel J, Castillo S, Traverse JP, et al. Ferric hydroxide oxide from the goethite process: Characterization and potential use. Industrial and Engineering Chemistry Research. 1993;**32**(9):1801-1804. DOI: 10.1021/ie00021a001

[11] Allan R, Haigh C, Hamdorf J. Method of Removing Dissolved Ferric Iron from Iron-Bearing Solutions. U.S. Patent 3781405DA.1973-12-25

[12] Han H, Sun W, Hu Y, et al. The application of zinc calcine as a neutralizing agent for the goethite process in zinc hydrometallurgy. Hydrometallurgy. 2014;**147-148**: 120-126. DOI: 10.1016/j.hydromet.2014.05.005

[13] Schlechtriem C, Ricci M, Focken U, et al. On a new method of removing iron from saline solutions as in the alum or sulphate of alumina manufactures from baunite, or other aluminous minerals. Journal of Chemical Technology & Biotechnology. 2010;**1**(7):275-285. DOI: 10.1002/jctb.5000010704

[14] Pelino M, Cantalini C, Abbruzzese C, et al. Treatment and recycling of goethite waste arising from the hydrometallurgy of zinc. Hydrometallurgy. 1995;**40**(1-2):25-35. DOI: 10.1016/0304-386X(95)00004-Z

[15] Reddy DHK, Yun YS. Spinel ferrite magnetic adsorbents: Alternative future materials for water purification? Coordination Chemistry Reviews. 2016;**315**:90-111. DOI: 10.1016/j.ccr.2016.01.012

[16] Reiss G, Hütten A. Magnetic nanoparticles: Applications beyond

data storage. Nature Materials. 2005;**4**(10):725-726. DOI: 10.1038/nmat1494

[17] Hokkanen S, Repo E, Lou S, et al. Removal of arsenic(V) by magnetic nanoparticle activated microfibrillated cellulose. Chemical Engineering Journal. 2015;**260**:886-894

[18] Ying A, Hou H, Liu S, et al. Ionic modified TBD supported on magnetic nanoparticles: A highly efficient and recoverable catalyst for organic transformations. ACS Sustainable Chemistry & Engineering. 2016;**4**(2):625-632. DOI: 10.1021/acssuschemeng.5b01757

[19] Tabakova T, Andreeva D, Andreev A, et al. Mechanism of the oxidative hydrolysis of iron(II) sulphate. Journal of Materials Science: Materials in Electronics. 1992;**3**(4):201-205. DOI: 10.1007/BF00703026

[20] Andreeva D, Mitov I, Tabakova T, et al. Formation of goethite by oxidative hydrolysis of iron(II) sulphate. Journal of Materials Science: Materials in Electronics. 1994;**5**(3):168-172. DOI: 10.1007/BF01198949

[21] Babcan J. Synthesis of jarosite $KFe_3(SO_4)_2(OH)_6$. Geology. 1971;**22**(2):299-304

[22] Claassen JO, Sandenbergh RF. Influence of mixing on the quality of iron precipitates in zinc-rich solutions. Hydrometallurgy. 2007;**87**(3-4):112-123. DOI: 10.1016/j. h ydromet.2007.02.002

[23] Yue T, Han H, Sun W, et al. Low-pH mediated goethite precipitation and nickel loss in nickel hydrometallurgy. Hydrometallurgy. 2016;**165**:238-243. DOI: 10.1016/j.hydromet.201 6.03.004

[24] Chang Y, Zhai X, Li B, et al. Removal of iron from acidic leach liquor of lateritic nickel ore by goethite precipitate. Hydrometallurgy. 2010;**101**(1-2):84-87. DOI: 10. 1016/j. hydrome t.2009.11.014

[25] Han H, Sun W, Hu Y, et al. Magnetic separation of iron precipitate from nickel sulfate solution by magnetic seeding. Hydrometallurgy. 2015;**156**:182-187. DOI: 10.1016/j. hydrome t.2015.07.001

[26] Kamimura Y, Lyoki K, Elangovan S, et al. OSDA-free synthesis of MTW-type zeolite from sodium aluminosilicate gels with zeolite beta seeds. Microporous and Mesoporous Materials. 2012;**163**:282-290. DOI: 10.1016/j. micromeso.2012.07.014

[27] Ran Z, Sun Y, Chang B, et al. Silica composite nanoparticles containing fluorescent solid core and mesoporous shell with different thickness as drug carrier. Journal of Colloid and Interface Science. 2013;**410**:94-101. DOI: 10.1016/j.jcis.2013.08.015

[28] Murphy CJ, Thompson LB, Chernak DJ, et al. Gold nanorod crystal growth: From seed-mediated synthesis to nanoscale sculpting. Current Opinion in Colloid & Interface Science. 2011;**16**(2):128-134. DOI: 10.1016/j. cocis.2011.01.001

[29] Wu Y, Ren X, Lu Y, et al. Crystallization and morphology of zeolite MCM-22 influenced by various conditions in the static hydrothermal synthesis. Microporous and Mesoporous Materials. 2008;**112**(1-3):138-146. DOI: 10.1016/j.micromeso.2007.09.022

[30] Han H, Sun W, Hu Y, et al. Induced crystallization of goethite precipitate from nickel sulfate solution by limonite seeding. Hydrometallurgy. 2017;**174**:253-257. DOI: 10.1016/j.hy dromet.2017.03.001

[31] Yue T, Xu Z, Hu Y, et al. Magnetic separation and recycling of goethite and calcium sulfate in zinc hydrometallurgy in the presence of maghemite fine

particles. ACS Sustainable Chemistry & Engineering. 2018;**6**(2):1532-1538. DOI: 10.1021/acssuschemeng. 7b03856

[32] Yue T, Sun W, Hu Y, et al. Mechanism of goethite precipitation on magnetite and maghemite nanoparticles studied by surface complexation/ precipitation modeling. Langmuir. 2018;**34**(50):15134-15142. DOI: 10.1021/ acs.langmuir.8b02571

[33] Cao S, Wang K, Zhou S, et al. Mechanism and effect of high-basicity chromium agent acting on Cr-wastewater-reuse system of leather industry. ACS Sustainable Chemistry & Engineering. 2018;**6**(3):3957-3963. DOI: 10.1021/acssuschemeng.7b04282

[34] Chrysochoou M, Dermatas D. Evaluation of ettringite and hydrocalumite formation for heavy metal immobilization: Literature review and experimental study. Journal of Hazardous Materials. 2006;**136**(1):20-33. DOI: 10.1016/j.jhazmat.200 5.11.008

[35] De Souza E, Silva PT, De Mello NT, et al. Extraction and recovery of chromium from electroplating sludge. Journal of Hazardous Materials. 2006;**128**(1):39-43. DOI: 10.1016/j.jha zmat.2005.07. 026

[36] Lin WY, Heng KS, Sun XL, et al. Influence of moisture content and temperature on degree of carbonation and the effect on Cu and Cr leaching from incineration bottom ash. Waste Management. 2015;**43**:264-272. DOI: 10.1016/j.wasman.2015.05.029

[37] Magalhães JM, Sliva JE, Castro FP, et al. Role of the mixing conditions and composition of galvanic sludges on the inertization process in clay-based ceramics. Journal of Hazardous Materials. 2004;**106**(2-3):169-176. DOI: 10.1016/j.jhazmat.2003.11.011

[38] Qian G, Yang X, Dong S, et al. Stabilization of chromium-bearing electroplating sludge with MSWI fly ash-based Friedel matrices. Journal of Hazardous Materials. 2009;**165**(1-3):955-960. DOI: 10.1016/j. jhazmat.2008.10.078

[39] Adhoum N, Monster L, Bellakhal N, et al. Treatment of electroplating wastewater containing Cu^{2+}, Zn^{2+} and Cr(VI) by electrocoagulation. Journal of Hazardous Materials. 2004;**112**(3):207-213. DOI: 10.1016/j.jhazmat.2004.04.018

[40] Kul M, Oskay KO. Separation and recovery of valuable metals from real mix electroplating wastewater by solvent extraction. Hydrometallurgy. 2015;**155**:153-160

[41] Hu J, Chen G, Lo IMC. Removal and recovery of Cr(VI) from wastewater by maghemite nanoparticles. Water Research. 2005;**39**(18):4528-4536. DOI: 10.1016/j.hydrome t.2015.04.021

[42] Miretzky P, Cirelli AF. Cr(VI) and Cr(III) removal from aqueous solution by raw and modified lignocellulosic materials: A review. Journal of Hazardous Materials. 2010;**180**(1-3):1-19. DOI: 10.1016/j.jhazmat.2010.04.060

[43] Suksabye P, Thiravetyan P. Cr(VI) adsorption from electroplating plating wastewater by chemically modified coir pith. Journal of Environmental Management. 2012;**102**:1-8. DOI: 10.1016/j.jenvman.2011.10.020

[44] Wen T, Wang J, Yu S, et al. Magnetic porous carbonaceous material produced from tea waste for efficient removal of As(V), Cr(VI), humic acid, and dyes. ACS Sustainable Chemistry & Engineering. 2017;**5**(5):4371-4380. DOI: 10.1021/acssuschemeng.7b00418

[45] Aoki T, Munemori M. Recovery of chromium(VI) from wastewaters with iron(III) hydroxide-I adsorption mechanism of chromium(VI) on iron(III) hydroxide. Water

Research. 1982;**16**(6):793-796. DOI: 10.1016/0043-1354(82)90006-9

[46] Erdem M, Tumen F. Chromium removal from aqueous solution by the ferrite process. Journal of Hazardous Materials. 2004;**109**(1-3):71-77. DOI: 10.1016/j.jhazmat.2004.02.031

[47] Sass BM, Rai D. Solubility of amorphous chromium(III)-iron(III) hydroxide solid solutions. Inorganic Chemistry. 1987;**26**(14):2228-2232. DOI: 10.1002/chin.198743013

[48] Ye Y, Jiang Z, et al. Efficient removal of Cr(III)-organic complexes from water using UV/Fe(III) system: Negligible Cr(VI) accumulation and mechanism. Water Research. 2017;**126**:172-178. DOI: 10.1016/j.watres.2017.09.021

[49] Yue T, Niu Z, Hu Y, et al. Cr(III) and Fe(II) recovery from the polymetallic leach solution of electroplating sludge by Cr(III)-Fe(III) coprecipitation on maghemite. Hydrometallurgy. 2019;**184**:132-139. DOI: 10.1016/j.hydromet.2018.11.013

[50] Nidheesh PV, Singh TSA. Arsenic removal by electrocoagulation process: Recent trends and removal mechanism. Chemosphere. 2017;**181**:418-432. DOI: 10.1016/j.chemosph ere.2017.04.082

[51] Luong VT, Cañas Kurz EE, Hellriegel U, et al. Iron-based subsurface arsenic removal technologies by aeration: A review of the current state and future prospects. Water Research. 2018;**133**:110-122. DOI: 10.1016/j.watres.2018.01.007

[52] Morales NA, Martínez D, García-Meza JV, et al. Total and bioaccessible arsenic and lead in soils impacted by mining exploitation of Fe-oxide-rich ore deposit at Cerro de Mercado. Durango, Mexico. Environment and Earth Science. 2015;**73**(7):3249-3261. DOI: 10.1007/s12665-014-3617-7

[53] Xiong Y, Tong Q, Shan W, et al. Arsenic transformation and adsorption by iron hydroxide/manganese dioxide doped straw activated carbon. Applied Surface Science. 2017;**416**:618-627. DOI: 10.1016/j.apsusc.2017.04.145

[54] Mohan D, Pittman CU. Arsenic removal from water/wastewater using adsorbents-A critical review. Journal of Hazardous Materials. 2007;**142**(1-2):1-53. DOI: 10.1016/j.jhazmat. 2007.01.006

[55] Luo T, Cui J, Hu S, et al. Arsenic removal and recovery from copper smelting wastewater using TiO_2. Environmental Science & Technology. 2010;**44**(23):9094-9098. DOI: 10.1021/es1024355

[56] Basha CA, Somasundaram M, Kannadasan T, et al. Heavy metals removal from copper smelting effluent using electrochemical filter press cells. Chemical Engineering Journal. 2011;**171**(2):563-571. DOI: 10.1016/j.cej.2011.04.031

[57] Ke Y, Shen C, Min XB, et al. Separation of Cu and As in Cu-As-containing filter cakes by Cu^{2+}-assisted acid leaching. Hydrometallurgy. 2017;**172**:45-50. DOI: 10.1016/j.hydrome t.2017.06.022

[58] Yue T, Niu Z, Hu Y, et al. Arsenic(V) adsorption on ferric oxyhydroxide gel at high alkalinity for securely recycling of arsenic-bearing copper slag. Applied Surface Science. 2019;**478**:213-220. DOI: 10.1016/j.apsusc.2019.01.249

[59] García-Sánchez A, Alastuey A, Querol X. Heavy metal adsorption by different minerals: Application to the remediation of polluted soils. Science of the Total Environment. 1999;**242**(1):179-188. DOI: 10.1016/S0048-9697(99)00383-6

[60] Donat R. The removal of uranium (VI) from aqueous solutions onto natural sepiolite. The Journal

of Chemical Thermodynamics. 2009;**41**(7):829-835. DOI: 10.1016/j.jct.2009.01.009

[61] Jiang W, Lv J, Luo L, et al. Arsenate and cadmium co-adsorption and co-precipitation on goethite. Journal of Hazardous Materials. 2013;**262**:55-63. DOI: 10.1016/j.jhazmat.2013.08.030

[62] His CKD, Langmuir D. Adsorption of uranyl onto ferric oxyhydroxides: Application of the surface complexation site-binding model. Geochimica et Cosmochimica Acta. 1985;**49**(9):1931-1941. DOI: 10.1016/0016-7037(85)90088-2

[63] Jeon KJ, Ahn JH. Evaluation of titanium tetrachloride and polytitanium tetrachloride to remove phosphorus from wastewater. Separation and Purification Technology. 2018;**197**:197-201. DOI: 10.1016/j.seppur.2018.01.016

[64] Lin CY, Nguyen MLT, Lay CH. Starch-containing textile wastewater treatment for biogas and microalgae biomass production. Journal of Cleaner Production. 2017;**168**:331-337. DOI: 10.1016/j.jclepro.2017.09.036

[65] Ménesguen A, Lacroix G. Modelling the marine eutrophication: A review. Science of the Total Environment. 2018;**636**:339-354. DOI: 10.1016/j.scitotenv.2018.04.183

[66] USEP Agency. Nutrient Criteria Technical Guidance Manual: Rivers and Streams. Washington: EPA; 2000

[67] Liu F, Yang J, Zuo J, et al. Graphene-supported nanoscale zero-valent iron: Removal of phosphorus from aqueous solution and mechanistic study. Journal of Environmental Sciences (China). 2014;**26**(8):1751-1762. DOI: 10.1016/j.jes.2014.06.016

[68] Guo J, Zhang Y, Zhao J, et al. Characterization of a bioflocculant from potato starch wastewater and its application in sludge dewatering. Applied Microbiology and Biotechnology. 2015;**99**(13):5429-5437. DOI: 10.1007/s00253-015-6567-4

[69] Vera L, Sun W, Iftikhar M, et al. LCA based comparative study of a microbial oil production starch wastewater treatment plant and its improvements with the combination of CHP system in Shandong, China. Resources, Conservation & Recycling. 2015;**96**:1-10. DOI: 10.1016/j.resconrec.2014.09.013

[70] Khlestkin VK, Peltek SE, Kolchanov NA. Review of direct chemical and biochemical transformations of starch. Carbohydrate Polymers. 2018;**181**:460-476. DOI: 10.1016/j.carbpo l.2017.10.035

[71] Cichy B, Kużdżał E, Krztoń H. Phosphorus recovery from acidic wastewater by hydroxyapatite precipitation. Journal of Environmental Management. 2019;**232**:421-427. DOI: 10.1016/j.jenvman.2018.11.072

[72] Johnston AE, Richards IR. Effectiveness of different precipitated phosphates as phosphorus sources for plants. Soil Use and Management. 2003;**19**(1):45-49. DOI: 10.1079/S UM2002162

[73] Chang Y, Yan X, Wang Q, et al. High efficiency and low cost preparation of size controlled starch nanoparticles through ultrasonic treatment and precipitation. Food Chemistry. 2017;**227**:369-375. DOI: 10.1016/j.foodchem.2017.01.111

[74] dos Santos JD, Veit MT, Juchen PT, et al. Use of different coagulants for cassava processing wastewater treatment. Journal of Environmental Chemical Engineering. 2018;**6**(2):1821-1827. DOI: 10.1016/j.jece.2018.02.039

[75] Verstraete W, Van de Caveye P, Diamantis V. Maximum use of

resources present in domestic "used water". Bioresource Technology. 2009;**100**(23):5537-5545. DOI: 10.1016/j.bior tech.2009.05.047

[76] Liu YN, Xue G, Yu SI. Comparing results of cultured and uncultured biological methods used in biological phosphorus removal. Journal of Environmental Sciences. 2007;**19**(11):1373-1379. DOI: 10.1016/S1001-0742(07)60224-1

[77] Luo D, Yuan L, Liu L, et al. The mechanism of biological phosphorus removal under anoxic-aerobic alternation condition with starch as sole carbon source and its biochemical pathway. Biochemical Engineering Journal. 2018;**132**:90-99. DOI: 10.1016/j.bej.2018.01.007

[78] Wang RM, Wang Y, Ma GP, et al. Efficiency of porous burnt-coke carrier on treatment of potato starch wastewater with an anaerobic-aerobic bioreactor. Chemical Engineering Journal. 2009;**148**(1):35-40. DOI: 10.1016/j.cej.2008.07.028

[79] Zou H, Wang Y. Phosphorus removal and recovery from domestic wastewater in a novel process of enhanced biological phosphorus removal coupled with crystallization. Bioresource Technology. 2016;**211**:87-92. DOI: 10.1016/j.biortech.2016.03.073

[80] Nie G, Wu L, Du Y, et al. Efficient removal of phosphate by a millimeter-sized nanocomposite of titanium oxides encapsulated in positively charged polymer. Chemical Engineering Journal. 2019;**360**:1128-1136. DOI: 10.1016/j.cej.2018.10.184

[81] Chen L, Li Y, Sun Y, et al. $La(OH)_3$ loaded magnetic mesoporous nanospheres with highly efficient phosphate removal properties and superior pH stability. Chemical Engineering Journal. 2019;**360**:342-348. DOI: 10.1016/j.cej.2018.11.234

[82] Lü C, Yan D, He J, et al. Environmental geochemistry significance of organic phosphorus: An insight from its adsorption on iron oxides. Applied Geochemistry. 2017;**84**:52-60. DOI: 10.1016/j.apgeochem.2017.05.026

[83] Zhang X, Lin X, He Y, et al. Adsorption of phosphorus from slaughterhouse wastewater by carboxymethyl konjac glucomannan loaded with lanthanum. International Journal of Biological Macromolecules. 2018;**119**:105-115. DOI: 10.1016/j.ijbiomac.2018.07.140

[84] Ye Y, Ngo HH, Guo W, et al. Insight into chemical phosphate recovery from municipal wastewater. The Science of the Total Environment. 2017;**576**:159-171. DOI: 10.1016/j.scitotenv.2016.10.078

[85] De-Bashan LE, Bashan Y. Recent advances in removing phosphorus from wastewater and its future use as fertilizer (1997-2003). Water Research. 2004;**38**(19):4222-4246. DOI: 10.1016/j.watres.2004.07.014

[86] Parsons SA, Smith JA. Phosphorus removal and recovery from municipal wastewaters. Elements. 2008;**4**(2):109-112. DOI: 10.2113/GSELEMENTS.4.2.109

[87] Pratt C, Parsons SA, Soares A, et al. Biologically and chemically mediated adsorption and precipitation of phosphorus from wastewater. Current Opinion in Biotechnology. 2012;**23**(6):890-896. DOI: 10.1016/j.copbio.2012.07.003

[88] Novais RM, Caetano APF, Seabra MP, et al. Extremely fast and efficient methylene blue adsorption using eco-friendly cork and paper waste-based activated carbon adsorbents. Journal of Cleaner Production. 2018;**197**:1137-1147. DOI: 10.1016/j.jclepro.2018.06.278

[89] Choi JM, Jeong D, Cho E, et al. Chemically functionalized silica gel with alkynyl terminated monolayers as an efficient new material for removal of mercury ions from water. Journal of Industrial and Engineering Chemistry. 2016;**35**:376-382. DOI: 10.1016/j.jiec.2016. 01.020

[90] Salama A, Aljohani HA, Shoueir KR. Oxidized cellulose reinforced silica gel: New hybrid for dye adsorption. Materials Letters. 2018;**230**:293-296. DOI: 10.1016/j.matlet.201 8.07.131

[91] Damtie MM, Kim B, Woo YC, et al. Membrane distillation for industrial wastewater treatment: Studying the effects of membrane parameters on the wetting performance. Chemosphere. 2018;**206**:793-801. DOI: 10.1016/j.chemosphere.2018.05.070

[92] Lv Y, Yan H, Yang B, et al. Bipolar membrane electrodialysis for the recycling of ammonium chloride wastewater: Membrane selection and process optimization. Chemical Engineering Research and Design. 2018;**138**:105-115. DOI: 10.1016/j.cherd.2018.08.014

[93] Zhang Y, Wei S, Hu Y, et al. Membrane technology in wastewater treatment enhanced by functional nanomaterials. Journal of Cleaner Production. 2018;**197**:339-348. DOI: 10.1016/j.jclepro.2018.06.211

[94] Yeap SP, Tia SY. Induced rapid magnetic sedimentation of stabilized-Fe_3O_4 nanoparticles by bridging and depletion flocculation. Chemical Engineering Research and Design. 2019;**142**:53-61. DOI: 10.1016/j.cherd.2018.12.004

[95] Tang H, Wang L, Sun W, et al. Electric arc furnace dust as magnetic carrier particles for removal of micro-fine particles from suspensions. Separation and Purification Technology. 2017;**176**:220-230. DOI: 10.1016/j.seppur.2016.12.024

[96] Lv M, Zhang Z, Zeng J, et al. Roles of magnetic particles in magnetic seeding coagulation-flocculation process for surface water treatment. Separation and Purification Technology. 2019;**212**:337-343. DOI: 10.1016/j.seppur.2018.11.011

[97] Ren W, Zhou Z, Jiang LM, et al. A cost-effective method for the treatment of reject water from sludge dewatering process using supernatant from sludge lime stabilization. Separation and Purification Technology. 2015;**142**:123-128. DOI: 10.1016/j.seppur. 2014.12.03 7

[98] Li B, Ma W, Wang H, et al. Optimization study on polymeric ferric sulfate (PFS) phosphorous depth removal. Applied Mechanics and Materials. 2011;**71-78**:3219-3223. DOI: 10.4028/www.scientific.net/AMM.71-78.3219

[99] Ruihua L, Lin Z, Tao T, et al. Phosphorus removal performance of acid mine drainage from wastewater. Journal of Hazardous Materials. 2011;**190**(1-3):669-676. DOI: 10.1016/j.jhazmat.2011.03.097

[100] Yan J, Jiang T, Yao T, et al. Preliminary investigation of phosphorus adsorption onto two types of iron oxide-organic matter complexes. Journal of Environmental Sciences (China). 2016;**42**:152-162. DOI: 10.1016/j.jes.2015.08.008

[101] Yang X, He J, Sun Z, et al. Effect of phosphate on heterogeneous Fenton oxidation of catechol by nano-Fe_3O_4: Inhibitor or stabilizer? Journal of Environmental Sciences (China). 2016;**39**:69-76. DOI: 10.1016/j.jes.2015.11.007

[102] Du C, Hu Y, Han H, et al. Magnetic separation of phosphate contaminants from starch wastewater using magnetic

seeding. The Science of the Total Environment. 2019;**695**:133723. DOI: 10.1016/j.scitotenv.2019.133723

[103] Du C, Yu J, Sun W, et al. Purification of starch and phosphorus wastewater using core-shell magnetic seeds prepared by sulfated roasting. Journal of Environmental Sciences (China). 2019;**81**:4-16. DOI: 10.1016/j.jes.2019.02.014

[104] Hemmelmann JC, Xu H, Krumm W. Empirical modeling of iron oxide dissolution in sulphuric and hydrochloric acid. Metallurgical and Materials Transactions B: Process Metallurgy and Materials Processing Science. 2013;**44**(5):1232-1235. DOI: 10.1007/s11663-013-9893-x

[105] Huang P, Deng S, Zhang Z, et al. A sustainable process to utilize ferrous sulfate waste from titanium oxide industry by reductive decomposition reaction with pyrite. Thermochimica Acta. 2015;**620**:18-27. DOI: 10.1016/j.tca.2015.10.004

[106] Ponomar VP, Dudchenko NO, Brik AB. Reduction roasting of hematite to magnetite using carbohydrates. International Journal of Mineral Processing. 2017;**164**:21-25. DOI: 10.1016/j.minpro.2017.05.005

Chapter 4

Advances in Sintering of Iron Ores and Concentrates

Jaroslav Legemza, Róbert Findorák, Mária Fröhlichová and Martina Džupková

Abstract

Chapter "Sintering of iron ores and concentrates" is focusing on the study of theoretical, thermodynamic and experimental results in the production of sinters from iron ores and concentrates. The authors of the chapter have long been interested with the production of sinter from iron ores and have recently also focused on the use of biomass as a substitute for a part of coke breeze in the production of iron sinter. Important characteristics of the chapter include the characteristics of iron ores and concentrates used to produce sinter including physico-chemical, mineralogical and metallurgical properties. Predicting the influence of the properties of iron ores and concentrates on the final quality of the sinter and on the production of pig iron is another part of the study. These properties are a key factor in achieving the highest possible agglomerate quality for pig iron production. The sintering process requires mathematical and physical modeling. For this reason, the authors created thermodynamic models of sintering including material-heat balance of sinter production. In the final part of chapter is the use of traditional and alternative carbonaceous fuels in the production of sinters, mainly in the context of replacement of coke breeze with biomass.

Keywords: iron ore, iron concentrate, sintering, coke, biomass, modeling

1. Introduction

Iron ores are very important not only in the production of pig iron in blast furnace, but it also in the production of sinter. These are mainly fine-grained iron ores and concentrates. The sinter is a basic input material for the production of pig iron in the blast furnace and plays an important role in the integrated metallurgical cycle. It is produced by high-temperature sintering of fine iron ore, iron ore concentrates and other ferriferous materials (e.g. secondary materials from iron and steel production). The main criterion of the sintering process is the quality produced sinter while maintaining the ecological nature of the production.

1.1 Characteristics of iron ores and concentrates used to produce sinter

1.1.1 Physical, chemical, mineralogical and metallurgical properties

Iron ore is very important for iron and steel industries. It is therefore elementary for the production of pig iron in blast furnace. Almost all (98%) iron ore is used in steelmaking. Iron ore is mined in about 55 countries. The five largest countries together produce about three quarters of world production, **Figure 1** [1]. Australia and Brazil dominate the world's iron ore production. Iron ores are classified according to the nature of the ore mineral and according to the nature of the gangue [1].

From a chemical point of view, iron ores are divided into groups: anhydrous oxides, hydrated oxides, carbonates and silicates. **Table 1** shows the classification of iron ores according to the type of ore mineral [2]. In practice, given this principle, only a few iron ore minerals out of a total of more than 300 types are considered. They are mainly oxide minerals, such as magnetite - Fe_3O_4 (72.36% Fe), hematite or martite (pseudomorphism of hematite after magnetite) - Fe_2O_3 (69.94% Fe), limonite - (62.85% Fe) (mixture of hydrated oxides Fe, mainly goethite with lepidocrocite $Fe_2O_3.nH_2O$, often with absorbed elements of vanadium, manganese, etc.). Carbonate ores based on siderite - $FeCO_3$ (48.30% Fe) can also be processed, as well as silicate ores (leptochlorites), e.g. chamosite. The highest natural iron content is in magnetite iron ores. Hematite ores are easily reducible. Chamosite is a

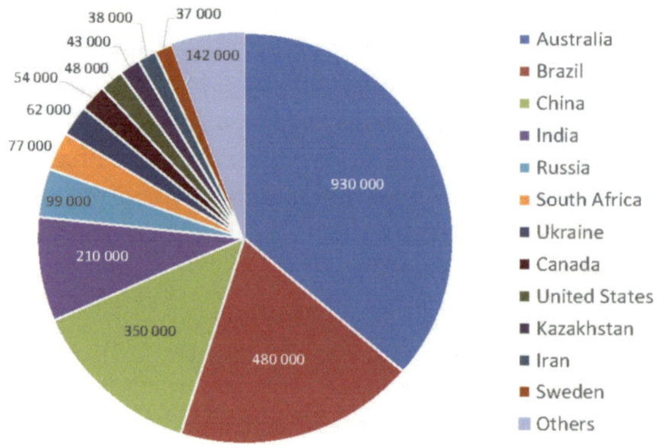

Figure 1.
Global production of usable iron ore (thousand metric tons) in 2019 [1].

A group of iron ores	Mineral	Chemical formula	Color	Density (g.cm^{-3})	Fe content in pure state (wt.%)
Magnetite ore	Magnetite	Fe_3O_4	Dark gray	5.17	72.4
Semimartite ore	Magnetite	Fe_3O_4	Black	5.1–5.2	—
Martite ore	Hematite	Fe_2O_3	Dadrk red	—	70.0
Hematite ore	Hematite	Fe_2O_3	Red	5.26	70.0
Limonite ore	Hydrohematite Goethite Lepidocrocite	$Fe_2O_3.nH_2O$	Dark red Light red Dark brown	3.1–4.4	63.0–69.0
Siderite ore	Siderite	$FeCO_3$	Gray	3.9	48.3
Chamosite ore	Chamosite	$Fe_4Al(Si_3AlO_{10}).(OH)_6.nH_2O$	Green Black	3.03–3.19	38.0

Table 1.
Classification of iron ores [2].

very difficult to reduce iron ore. Oxidation of magnetite creates varieties of hematite-martite and semimartite (according to the degree of oxidation). Most of the world's production of pig iron - 90%, is produced from iron ores of an oxidic nature, of which 5% from magnetite ores and 85% from hematite ores.

Figure 2 shows some known iron ores from world. The samples come from the Atlas of iron ores, which was created at the VUHŽ Dobrá (Czech republic) research institute [3] and with which the authors of this chapter have collaborated in the past.

The evaluation of the properties of iron ores shows that the best ores are from Brazil, Australia and Venezuela, **Table 2** [3]. They have a suitable particle size distribution, excellent chemical and mineralogical composition, they are well reducible and stable after temperature tests. The optimal piece size of raw iron ore for blast furnaces is 10–40 mm and it is necessary to completely exclude dust fractions below 5 mm. Kryvyi Rih ore has an unsatisfactory chemical composition - low Fe content and high SiO_2 content, which is unsuitable for the blast furnace process. On the other hand, this ore has good physical properties (eg. strength and stability) and metallurgical properties (reducibility and plasticity). Ore Algeria has a high proportion of large grains (above 40 mm) and lower strength, reducibility is average and plasticity is unsuitable (high temperature range). Indian ore has worse metallurgical properties - lower reducibility and higher plasticity temperature range. The softening onset temperature and the softening interval characterize the plasticity and also affect the position and height of the plastic zone in the blast furnace. The course of softening depends on the type of ore and cannot be influenced. The blast furnace charge should therefore contain a minimum number of different iron ore raw materials.

Sinter grade ores and concentrates are characterized as iron ore raw materials with the required granulometry, composition and properties. The following

Figure 2.
Iron ores (the photographs were created from atlas of ores [3] by the authors of this chapter).

Iron ore							
		Brazil MBR	**Kryvyi Rih (H)**	**Australia**	**Algeria**	**Venezuela**	**India**
Analysis (wt.%)	Fe	68.13	47.59	67.20	59.50	65.09	65.50
	FeO	1.12	2.00	0.49	0.24	1.08	0.53
	Fe_2O_3	96.16	65.82	95.50	84.81	91.81	93.08
	Mn	0.04	0.05	0.05	1.28	0.26	0.08
	SiO_2	0.53	24.30	1.77	2.85	4.49	2.81
	Al_2O_3	1.61	3.25	1.33	0.65	1.14	2.07
	CaO	0.23	0.22	0.28	1.31	0.15	0.35
	MgO	0.03	0.20	0.10	0.72	0.19	0.06
	P	0.05	0.05	0.02	0.02	0.03	0.03
	S	0.01	0.03	0.01	0.03	0.01	0.02
	Na_2O	0.07	0.14	0.01	0.08	0.07	0.09
	K_2O	0.04	0.95	0.02	0.06	0.06	0.07
Grain (mm)	< 1	1.55	2.74	1.92	0	8.03	6.34
	< 5	5.64	4.75	5.85	0	13.22	10.30
	> 10	81.13	91.40	70.54	98.70	64.16	85.04
	> 25	30.18	49.54	14.03	97.15	0.80	47.18
	> 40	5.36	23.04	0	84.19	0	16.80
	d_A	20.46	28.60	16.30	59.85	11.89	25.84
Apparent density ($g.cm^{-3}$)	ρ_A	1.99	1.78	2.22	1.45	2.42	2.15
Real density ($g.cm^{-3}$)	ρ	5.21	3.85	4.82	4.11	4.95	4.73
Porosity (%)	P	18.6	7.8	16.2	33.6	12.1	20.9
Surface ($m^2.g^{-1}$)	S	1.22	0.96	1.63	5.00	0.44	6.55
Drum strenght (%)	+ 6.3 mm	78.99	84.63	89.42	75.12	88.94	77.62
Heat test (%)	+ 6.3 mm	76.62	73.04	69.42	85.70	69.32	62.26
Reducibility (%)	R_i	40	35	42	33	43	30
Reducibility (min)	R_{i60}	101	110	95	121	100	170
Plasticity (°C)	Pl_{60}	1174–1320	1150–1483	1380–1423	990–1390	1265–1490	1210–1480
Mineralogy	XRD	Hematite, goethite	Martite, hydrogoethite quartz	Hematite, goethite, quartz	Limonite, goethite, hydrogoethite, quartz	Hematite, martite, goethite, quartz	Martite, goethite, hydrogoethite, quartz

Reducibility is according ISO 7992.

Table 2.
Properties of iron ores (according [3]).

Tables 3, 4 provide a basic chemical analysis of some, also used in Slovakia, sinter grade ores and concentrates for the production of sinter [4].

The authors of this chapter used in their research mainly such iron-bearing raw materials as sinter grade ores supplied from Kryvyi Rih and Brazil and concentrates from Michailovsky and Inguletsky MBCs (Mining-Benefeciation Combines), **Figure 3**. The grain size of the sinter grade ores is 90% below 10 mm, the grain size of the iron concentrates is 90% below 0.04 mm. Iron ores and concentrates before sintering are pretreated in the granulation process. The values of specific surface of the granulated materials should not be lower than 2000 $cm^2.g^{-1}$, which corresponds to 100% portion of the size fraction below 0.15 mm, from which 75% of the size fraction below 0.04 mm.

Table 5 shows the chemical composition of iron-bearing raw materials, where the difference in the richness of ores can be seen. **Figure 4** shows the structure of the samples and Energy Dispersive X-Ray Analysis (EDX), **Table 6**. The larger

Iron ore	Fe	Mn	SiO_2	Al_2O_3	CaO	MgO	P	S	Na_2O	K_2O	Zn	H_2O
	(wt.%)											
Krivbas	64.04	0.02	6.24	0.68	0.05	0.05	0.035	0.014	0.183	0.046	0.002	3.9
Sucha Balka	59.86	0.02	12.10	0.77	0.06	0.14	0.026	0.011	0.148	0.043	0.004	3.8
Zaporozska	62.70	0.06	7.38	0.84	0.52	0.23	0.023	0.020	0.055	0.027	0.004	4.6
Brazil MBR	65.00	0.08	3.50	0.90	0.08	0.06	0.042	0.005	0.005	0.008	0.003	7.4
CVG-Venezuela	65.66	0.31	1.16	0.89	0.10	0.05	0.074	0.014	0.013	0.024	0.003	6.8
Liberia	64.50	0.16	6.73	0.96	0.10	0.05	0.047	0.004	0.013	0.015	0.003	7.5
CIL Trinidad	67.57	0.01	1.48	0.47	0.70	0.52	0.001	0.003	0.001	0.001	0.001	4.0
Hope Downs	64.79	0.02	1.36	0.77	0.03	0.07	0.057	0.007	0.016	0.001	0.006	5.5

Table 3.
Chemical analysis of selected sinter grade ores in the delivered state [4].

Iron concentrate	Fe	Mn	SiO_2	Al_2O_3	CaO	MgO	P	S	Na_2O	K_2O	Zn	H_2O
	(wt.%)											
Southern MBC	67.50	0.023	5.90	0.15	0.12	0.35	0.010	0.027	0.059	0.014	0.003	10.2
Central MBC	64.90	0.032	8.49	0.31	0.19	0.35	0.015	0.073	0.038	0.042	0.003	9.3
Stoilensky MBC	66.36	0.041	6.78	0.18	0.21	0.44	0.014	0.035	0.045	0.049	0.003	8.8
Inguletsky MBC	65.68	0.027	7.28	0.25	0.23	0.29	0.013	0.066	0.032	0.042	0.003	9.4
Lebedinsky MBC	67.79	0.025	4.92	0.14	0.16	0.33	0.011	0.044	0.062	0.035	0.002	9.5
Kovdor MBC	64.74	0.430	0.42	1.78	0.23	5.49	0.038	0.269	0.026	0.032	0.034	2.7
LKAB magnetic fines	70.78	0.050	0.61	0.25	0.16	0.33	0.021	0.008	0.041	0.045	0.003	2.7
Venezuela	69.65	0.056	0.97	0.78	0.24	0.13	0.075	0.020	0.011	0.022	0.005	7.0

Table 4.
Chemical analysis of selected iron ore concentrates, as delivered [4].

Figure 3.
Iron raw materials for sintering process.

Iron ore material	Fe	FeO	Fe_2O_3	Mn	SiO_2	Al_2O_3	CaO	MgO	P	S	Na_2O	K_2O
	(wt.%)											
Sinter grade ore Krivbas	57.86	0.72	81.94	0.05	15.17	0.74	0.15	0.25	0.05	0.01	0.08	0.05
Sinter grade ore Brazil	65.23	0.14	93.12	0.57	2.21	1.32	0.10	0.10	0.05	0.01	0.05	0.05
Concentrate Inguletsky MBC	67.91	28.45	65.57	0.07	4.92	0.18	0.59	0.49	0.02	0.09	0.06	0.07

Table 5.
Chemical analysis of selected iron ore materials, as delivered.

grains of Brazil sinter grade ore are practically free of impurities and have a relatively homogeneous structure. In addition to iron oxides, the smaller sinter grade ore Brazil grains also contain impurities in the form of silicon and aluminum oxides.

In general, the richness of concentrates is in the range of 65–70%, while sinter grade ores have this interval wider and shifted slightly lower (55–67%).

Taking a closer look at the chemical composition of iron commodities, we can see a fundamental difference in the FeO content of sinter grade ores and concentrates (**Table 5**), which is related to the enrichment processes of concentrate production and mainly the hematite character of the sinter grade ores used, **Table 7**.

The SiO_2 content (in the form of quartz) is desirable at a minimum level and depends on the specific ore resp. concentrate. Manganese is also a welcome ingredient and increases the utility value of ore along with titanium and vanadium. Undesirable impurities are mainly sulfur, phosphorus, zinc, lead, arsenic, copper, sodium, potassium, which are chemically bound in the minerals of the ore part of the burden, as well as in the agglomeration fuel.

The form of occurence of harmful elements in iron ores is as follows [5–7]:

- Arsenic is present in arsenopyrite or arsenolite.
- Zinc is present in the form of smithsonite or sphalerite.
- Lead occurs in the form of galena.
- Copper can be present in the form of chalcopyrite or chalcosine.
- Phosphorus forms phosphite, which is a part of apatite.
- Sulfur is present in the form of sulphides such as FeS_2 and FeS and of sulphates such as $CaSO_4$, $MgSO_4$, $BaSO_4$.

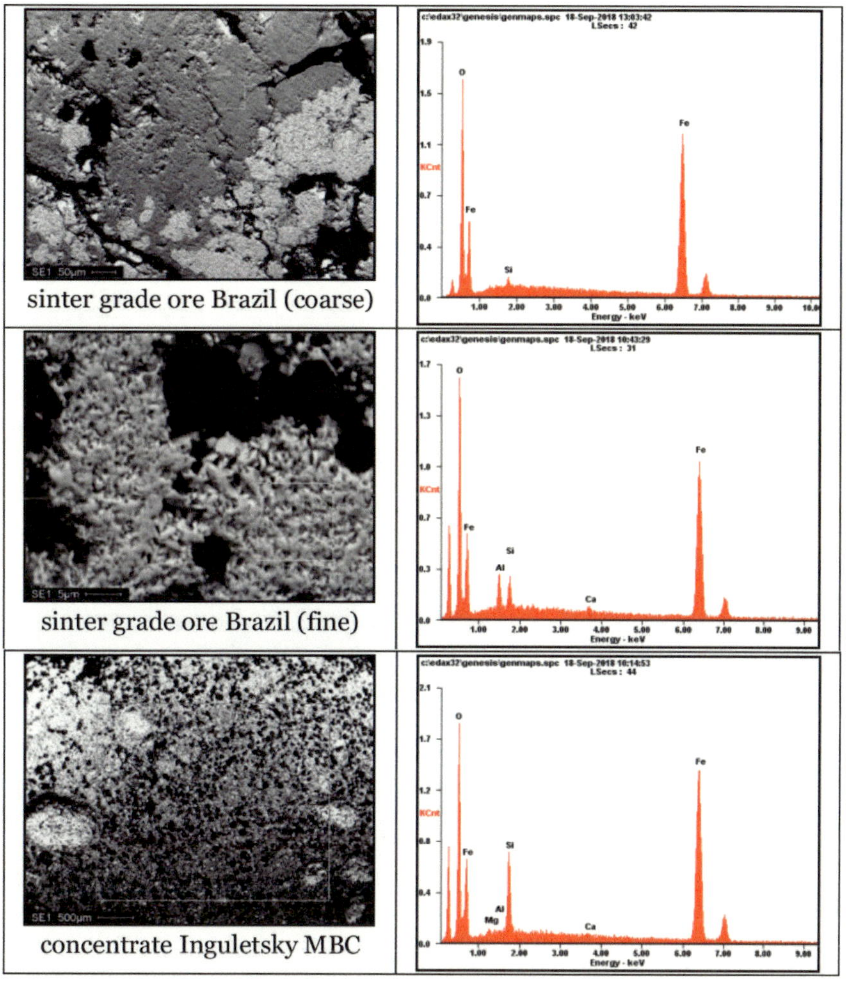

Figure 4.
EDX analysis of Brazil iron ore and Inguletsky iron concentrate.

Element	Sinter grade ore Brazil (coarse)		Sinter grade ore Brazil (fine)		Concentrate Inguletsky MBC	
	Wt%	Atom %	Wt%	Atom %	Wt%	Atom %
O	18.45	44.13	17.56	41.15	14.49	35.33
Fe	81.55	55.87	75.40	50.62	78.21	54.61
Mn	—	—	1.56	1.06	—	—
Si	—	—	3.00	4.01	5.77	8.02
Al	—	—	1.85	2.57	0.25	0.36
Ca	—	—	0.64	0.59	0.57	0.55
Mg	—	—	—	—	0.71	1.14

Table 6.
EDX analysis of selected iron ore materials.

Quality sinter grade ores is characterized by a high content of total iron (min. 63%), Al_2O_3 content max. up to 1.3%, SiO_2 up to 6%, P below 0.04% and alkali content max. up to 0.08% [8].

Mineralogical phase	Content of phase (wt%)		
	Sinter grade ore Krivbas	Sinter grade ore Brazil	Concentrate Inguletsky MBC
Hematite	72.50	75.48	—
Magnetite	—	—	89.70
Goethite	15.20	18.73	—
Mayenite	—	1.23	—
Quartz	12.30	4.56	10.30

Table 7.
RTG analysis of selected iron raw materials.

From the mineralogical point of view, it is important to know the ore texture and structure that characterizes the distribution of individual mineralogical components, respectively indicates the shape and size of the mineral components and the nature of their structure. The mineralogical composition of Fe ores is characterized by the majority shape of the grain in terms of their structure, **Table 8** [2].

Some iron bearing materials were evaluated prior to laboratory experiments and the grain shape factor was determined [9]. The grain shape factor specifies calculation a grain periphery, which differs from the circle (for circle is shape factor = 1). The iron ore shown in **Figure 5a** consists the grains with polyhedron shape (shape factor = 0.59) in a size of 5–40 μm. The iron ore shown in **Figure 5b** consists of the grains with polyhedron shape (shape factor = 0.94) with smooth surfaces in a size of 10–50 μm. In this case, the hematite grains are clearly visible. The iron ore shown in **Figure 5c** consists of the grains with partially smoothened edges of polyhedron shape (shape factor = 0.78) in a size of 10–100 μm. The iron ore shown in **Figure 5d** consists of the grains with lamellas shape (shape factor = 0.70) in size of 10–250 μm. The hematite grain can be seen at the top.

In addition to the chemical-mineralogical composition of sinter grade ores and concentrates, it is also necessary to know their granulometry. The grain size

Mineralogical composition	Shape of grain
Magnetite, pyrite, quartz	Spherical, cubic
Magnetite, hematite, goethite, quartz	Cubic, platelet-shaped
Hematite, goethite, quartz, kaolinite	Platelet-shaped, fragmentary
Hematite	Polyhydral, plate-shaped

Table 8.
Major grain shape types of iron ores [2].

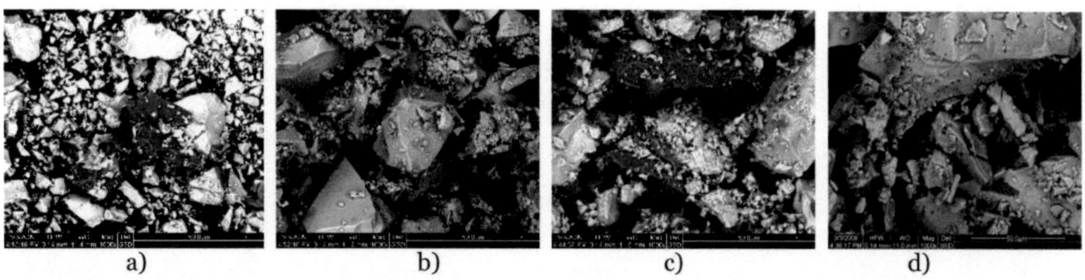

Figure 5.
Microstructures of iron ores with various shape factors [9]. (a) 0.59, (b) 0.94, (c) 0.78, (d) 0.70.

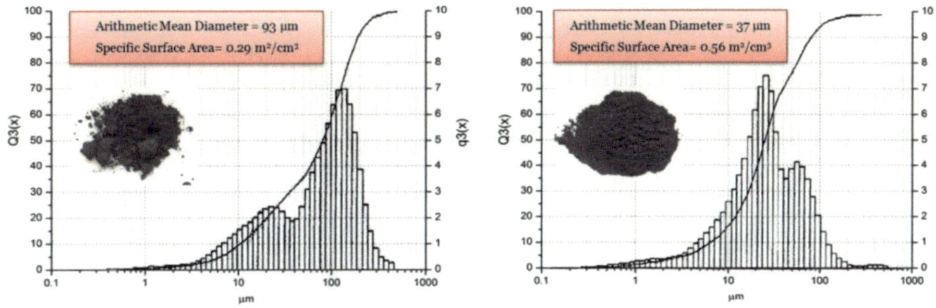

Figure 6.
Distribution of grain in iron concentrates. (a) Concentrate Kovdor MBC, (b) concentrate Lebedinsky MBC.

distribution will be related to the processing and treatment of extracted ore or the production of iron concentrate. Fine-grained concentrates are the product of flotation enrichment of ores and their predominant grain size is about 0.04 mm. The granularity is also directly related to the mineralogical composition and structure of the ore. The concentrates used in the sintering process have an overall wide grain size range. Most Ukrainian and Russian concentrates are very fine-grained in nature, where the extracted magnetite quartzite is crushed and ground below 0.075 mm and enriched by magnetic separation into a concentrate with Fe content up to 68%. Swedish magnetite concentrate, on the other hand, has only about 4% of the fraction below 0.1 mm at a richness of about 70%. Canadian magnetite concentrates have a proportion below 0.1 mm of about 15%, while the upper grain limit does not exceed 1 mm. The results of analysis of the selected concentrates are given in **Figure 6**.

1.1.2 The influence of the properties of iron ores and concentrates on the final quality of the sinter and on the production of pig iron

The operation of the blast furnace and the results of its work are most often evaluated according to the output and consumption of coke. Changes in the chemical composition and particle size distribution of the iron-bearing materials significantly affect their technological properties and thus the balance of components and the course of the blast furnace process. The development of iron metallurgy is conditioned by the quantity and quality of iron ores. The raw material base for iron production is characterized:

- lack of high-quality natural ores - only 10–12% of the world's iron ore reserves meet the current requirements of blast furnace practice,
- uneven distribution of world stocks in individual countries,
- low content of metal-bearing substance in mined ore,
- use of high-performance equipment in ore mining and enrichment.

To meet the requirements of the metallurgical industry, mined ores are increasingly treated and processed before being used in blast furnaces, and the treatment of ores must ensure:

- an increase in the iron content and removal of harmful and unwanted impurities,

- improving the physical and mechanical properties of ores,
- improving the reducibility of individual types of ores,
- averaging and stability of chemical composition and lumpiness of iron ore raw materials.

The implementation of the above requirements will achieve good permeability of the burden column, reduction of the amount of slag-forming additives in the burden, reduction of specific coke consumption, increase of blast furnace output, improvement of pig iron quality, more even operation of the furnace without more serious failures and fluctuations, reduction of iron production costs. In the blast furnace process, zinc and lead from iron raw materials belong among the so-called harmful elements with a significant influence on the formation and growth of sediments [5–7]. Potassium and sodium are undesirable in the blast furnace charge due to the disruption of the integrity of the carbon and graphite linings [5, 7].

The requirements for the quality of sinter are constantly increasing. The chemical and mineralogical composition of sintered materials affects the wettability of their surfaces and also influences considerably the strength of the binding of the material grains in the green pellets, and in this way also influences the conditions of sintering. The sintering rate depends on the initial properties of the iron ores and concentrates and the heating conditions of the sintered materials, i.e. it depends on the geometric parameters of ores and concentrates (grain size composition, the size of specific surface, grain morphology, microgeometry of the surface, porosity) and on the structural activity of the material.

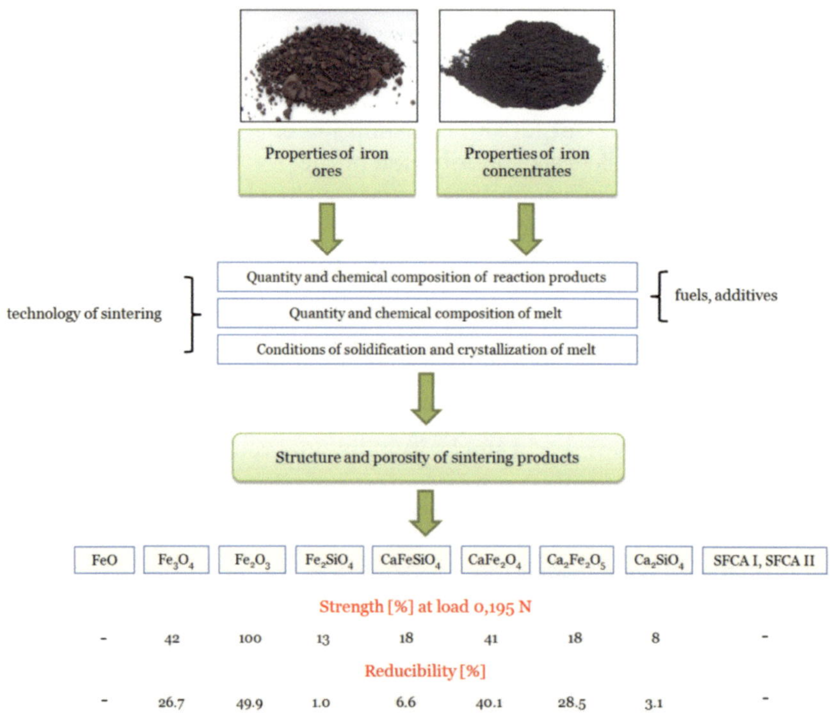

Figure 7.
The influence of various factors on the Fe sinter (modified by authors according [2]). SFCA = silicoferrites of calcium and aluminum, strength and reducibility were realized on pure mineral compounds.

A good quality sinter is characterized by a suitable iron content, high reducibility, good strength and low fine grain shares content prior to charging into blast furnace and high strength after reduction in the blast furnace shaft. The influence of the properties of iron ores and concentrates on the final quality of the sinter is specified on **Figure 7**. **Figure 7** shows very good mechanical (strength) and metallurgical (reducibility) properties of major oxides (hematite and magnetite) and calcium ferrites. On the other hand, silicates have unfavorable properties.

The sintering product should have the proper physical features to bear transporting and should not produce dust while in the blast furnace. The sinter should be of good chemical, mineralogical and metallurgical properties and should contain as few detrimental admixtures as possible. In **Table 9**, some important properties of industrially produced sinters are listed, while the critical requirement for all sinter properties is stability [10].

Property of Fe sinter	SI unit	Min	Max
Content of Fe_{TOT}	[%]	48.20	57.00
Content of FeO	[%]	8.70	19.80
Content of CaO	[%]	7.30	14.40
Content of SiO_2	[%]	5.40	9.60
Content of P	[%]	0.02	0.04
Content of S	[%]	0.03	0.05
Content of $Na_2O + K_2O$	[%]	0.05	0.08
Basicity	[−]	1.15	2.10
Granulometry	[mm]	5.00	50.00
Porosity	[%]	27.00	38.00
Real density	[$g.cm^{-3}$]	4.13	4.44
Reducibility (ISO 7992)	[%]	60.00	85.00
Reducibility (ISO 4695)	[%/min]	0.80	1.40
Drum strength +6.3 mm	[%]	65.00	78.00
Abrasion index −0.5 mm	[%]	4.20	9.80

Table 9.
Properties of Fe sinters [10].

1.2 Modeling and simulation of sinter production under laboratory conditions

1.2.1 Material-heat balance and thermodynamic study of sinter production

The course of processes in the sintered material can be evaluated on the basis of changes in the physical and metallurgical properties of the sintering product. The transformation of the components of a material is related to the decrease in Gibbs free energy. It is possible to calculate the maximum work of reactions, i.e. oxidation reactions, reaction in solid state, reactions during formation of a melt and also reactions taking place during the recrystallization and cooling down of sintering products.

Thermodynamic calculations are thus essential when determining the characteristics of a technological process and they enable one to clarify the formation of the major sintering products. By changing the basic condition of the thermodynamic system, it is feasible to find the optimum operating conditions of the sintering process and minimize the consumption of raw materials and energy [11].

For mathematical modeling the basic chemical reactions with standard Gibbs energy and mass and thermal balance were calculated. Thermodynamic data was obtained from the software HSC Chemistry. HSC Chemistry offers powerful calculation methods for studying the effects of different variables on the chemical system at equilibrium. The aim is to get the simplest approach (using this software to calculate equilibrium) which allows one to predict the output parameters (amounts, chemistry, mineralogical composition, and total heat) based on the initial composition analysis [11].

The mathematical model used by the authors in modeling the sintering process allows the calculation of Gibbs equilibrium diagrams, which characterize the change in the equilibrium composition of reactants and reaction products with changes in temperature. Using Kellogs diagrams of stability areas in the considered systems based on the combustion of carbonaceous fuel in the sintering charge, it is also possible to determine the stability of individual phases at different partial pressures of gaseous components. The modeled systems and the results of thermodynamic calculations can specify the influence of the amount and type of fuel used on oxidation–reduction processes. The said model allows the control of the overall thermal effect of the sintering process during individual instances of experimental laboratory sintering while it employed the prediction of sinter phase composition at the sintering temperatures in the calculations. Since the phase composition of the sinter is determined on samples of sinter after cooling, the computational model should bring a new perspective of the processes taking place during sintering [11].

Figure 8 shows the global method with modeling the mass and thermal balance of the sintering process. Pursuant to the modeling of the sintering process in the

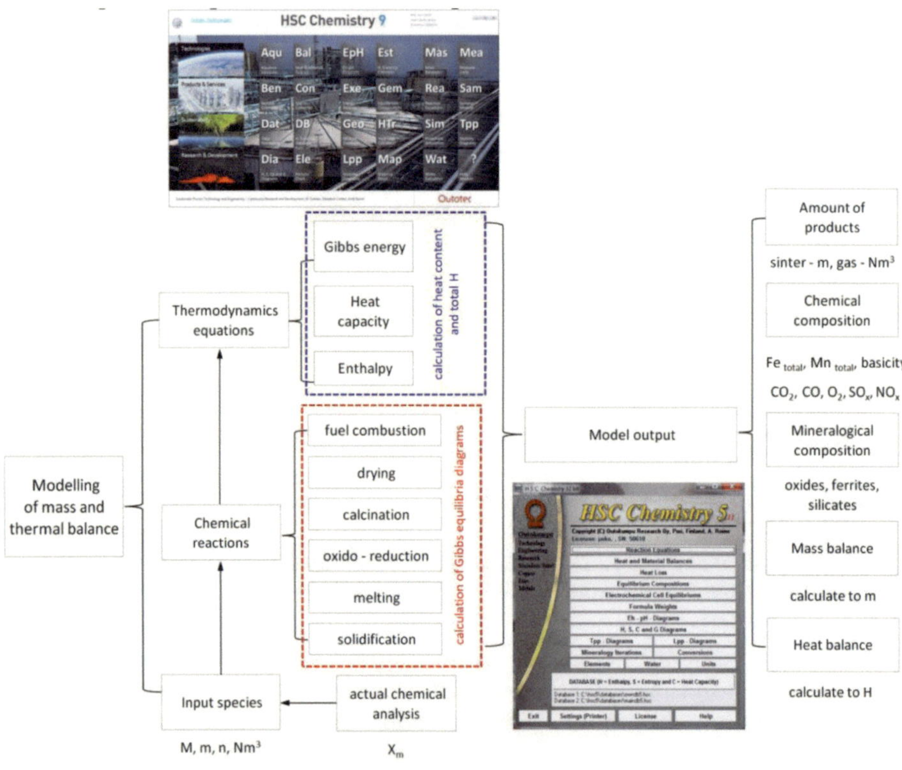

Figure 8.
Scheme presenting the global method with modeling the mass and thermal balance (modified by authors according [11]).

laboratory conditions, it was feasible to specify the created model by the following parameters [11]:

- calculation of the quantity and chemical composition of sinter,
- prediction of the mineralogical composition of sinter at the sintering temperatures,
- calculation of the mass and thermal balance on the basis of the input and output enthalpies of individual components.

The authors of this chapter have already carried out a large number of laboratory experiments [9–11], the results of which can be compared with the calculated thermodynamic models. It is apparent that the model calculations of added fuel in charge are highly correlated with the experimentally determined values, **Figure 9a**. A higher correlation was found for the yield of produced sinter (calculation for 100 kg of charge), **Figure 9b**.

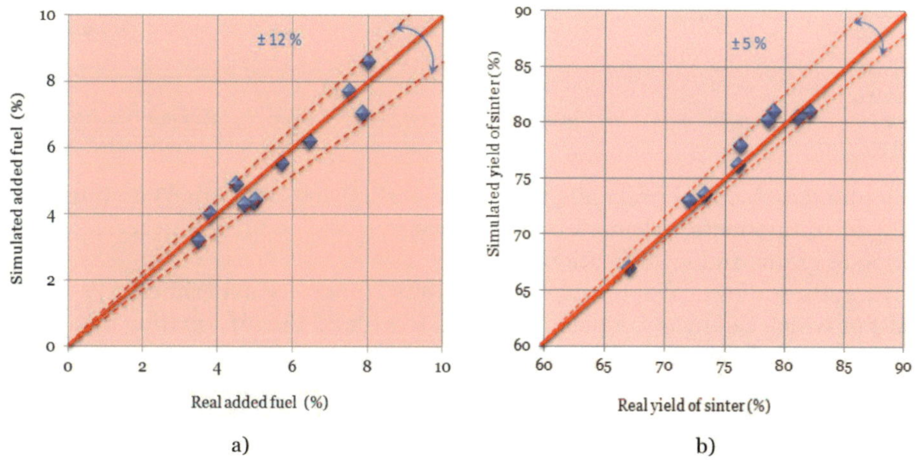

Figure 9.
Comparison between the real and simulated added fuel for sintering (a) and real and simulated amount of the sinter (b).

1.2.2 Specification of the model of laboratory sintering pot and monitoring of the sintering process

The sintering process is thermal process used to transform fine particles of iron ore and concentrate into porous product known as sinter. In the sintering process fuel is in the form of coke breeze (or in the form various types of carbon fuels) used for production of iron-ore sinter [10]. In this sintering process are basic materials mixed, granulated, ignited and fired at a temperatures 1200–1380°C. Sintering of materials can take place under the temperature conditions, which allow binding the particles by reaction in solid state, or under the temperature conditions, which allow the origin of the melt acting as a binding phase after its cooling. In the case of sintering with the presence of the melt, its quantity and chemical compositions is important.

The process of laboratory sinter production is divided into two stages – cold section and hot section, **Figure 10**. In the cold section, the supply of raw materials, adjustment of grain size to the required piece size, averaging of the chemical composition of materials and granulation of the final sintering burden is ensured. The hot section ensures ignition of the sintering charge, sintering of the sintering charge, removal of hot flue gases (subsequent cleaning) and cooling of the sinter.

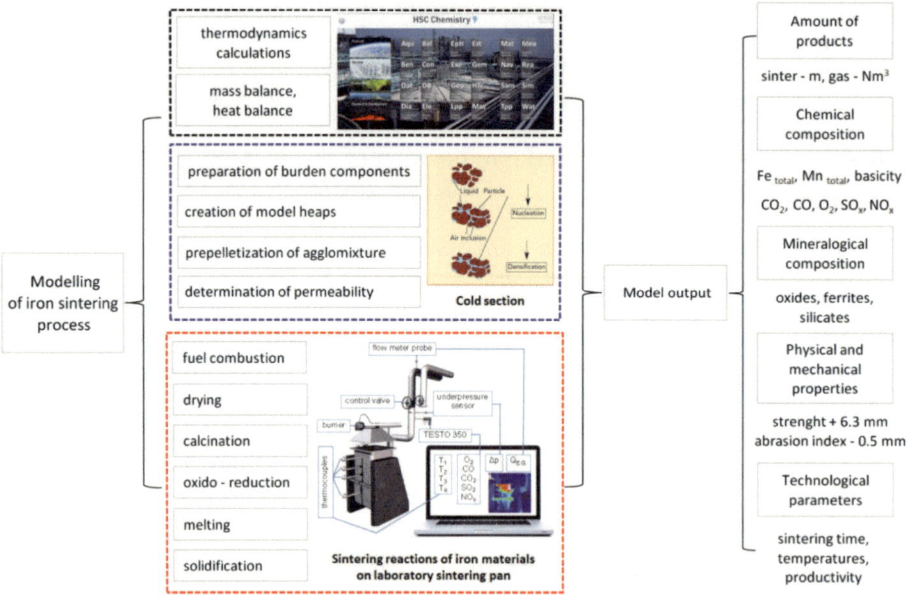

Figure 10.
Scheme presenting the global method with modeling of the iron sintering process in laboratory conditions in Slovakia.

Many physico-chemical processes take place during sintering – fuel combustion, drying of components, calcination of carbonates, oxido-reduction processes, melting of grains and solidification.

The authors of this chapter carried out many laboratory experiments [9–13], the results of which can be generalized and used to expand the information database on the use of carbonaceous fuels in the production of sinter. These experiments were carried out in a laboratory sintering pot (LSP), which is located at the Institute of Metallurgy, Faculty of Materials, Metallurgy and Recycling, Technical University of Košice in Slovakia, **Figure 10**.

In order to simulate the production of sinter close monitoring of the sintering process is necessary. Laboratory sintering pot is fully equipped with measuring devices and analyzers. The temperature was measured by thermocouples. For the high temperature range in the sintered layer, three thermocouples of the PtRh10-Pt type were used. The flue gas temperature was read at two levels by NiCr-Ni type thermocouple. Chemical composition and temperature of the flue gas were analyzed by TESTO 350 device. The differential pressure was measured by Annubar type probe, which served for calculating the amount of sucked air (or flue gas). All quantities were read at 15-second intervals and collected in a logger. After each experiment, the collected data were transformed into a form usable on a personal computer.

Implementation of an experimental sintering model is also possible using the monitoring of the temperature field of the sintering pot by a thermal imaging camera. With the help of the sensed thermal imaging profile of the sintering pot (**Figure 11**), it is possible to monitor the displacement of the fuel combustion zone in the sintering layer. By sensing the surface of the layer after ignition of the burden, (**Figure 12**), it is possible to monitor the inhomogeneity of the temperature field of the sintered layer. The thermal profile of thermal imaging monitoring indicates different conditions of fuel combustion in the volume of the layer, which may be caused by heterogeneity of fuel distribution. The model of LSP has been innovated by a transparent high-temperature wall that allows visual monitoring of

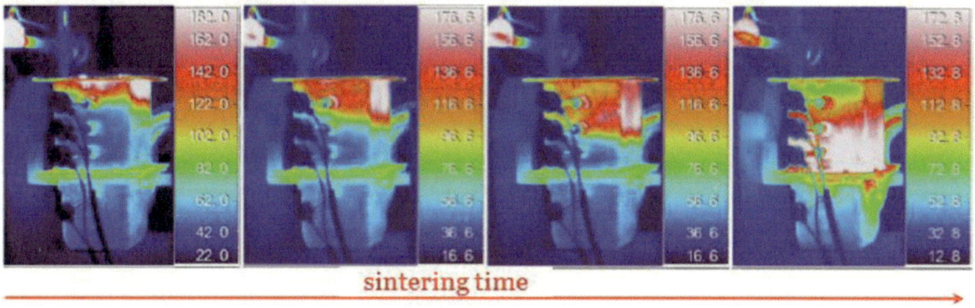

Figure 11.
Thermal imaging of a laboratory sintering pot.

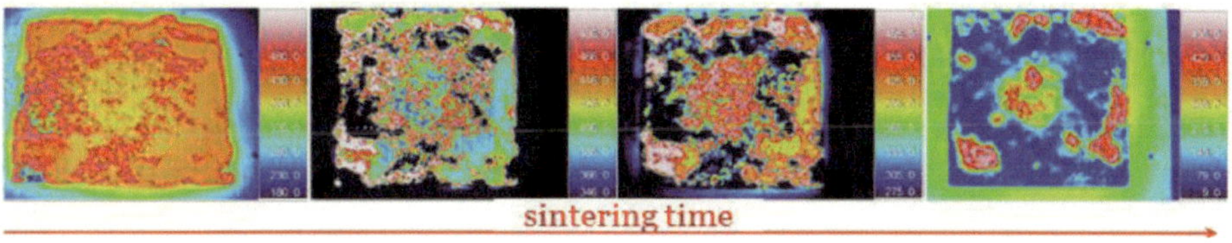

Figure 12.
Thermal imaging of the surface of the burden after ignition.

Figure 13.
Course of sintering and moving of sintering zone.

the combustion zone in the sintered layer during the production of iron ore sinter, **Figure 13**.

1.2.3 Technological and ecological aspects of the production of sinters from poor and rich iron-bearing materials

As part of the implementation of laboratory experiments on a laboratory sintering pot, the following iron-bearing raw materials were used for sintering, **Table 10**.

In the sintering process, standard coke breeze was used as fuel. These raw materials were incorporated in the prepared sintering mixtures, which had a basicity in the range of 1.7–2.8. The next relation was used to calculate the basicity: $B = (wt.\%CaO + wt.\%MgO)/(wt.\%SiO_2 + wt.\%Al_2O_3)$. It was therefore the production of highly basic sinters. The produced sinters had content of iron in the range of about 46–52%.

In the experiments, the ratio of ferriferous raw materials (Krivbas and Michajlov) – 100% sinter grade ore and 100% concentrate was changed. Concentrate Nižná Slaná was used in mixtures of sinter grade ore/concentrate. In this chapter, primarily the experiments with separate ferriferous raw materials are specified due to the more significant impact of fuel consumption on sinter quality.

Iron ore material	Fe	FeO	Fe_2O_3	Mn	SiO_2	Al_2O_3	CaO	MgO	P	S	Na_2O	K_2O
	(wt.%)											
Sinter grade ore Krivbas	60.70	0.14	86.26	0.02	9.61	1.35	0.07	0.06	0.05	0.02	0.29	0.08
Concentrate Michajlov	64.52	23.57	66.44	0.01	8.60	0.06	0.19	0.09	0.01	0.01	0.08	0.20
Concentrate Nižná Slaná	51.76	20.02	45.19	1.97	5.30	1.83	3.82	6.58	0.08	0.12	0.07	0.42

Table 10.
Chemical analysis of iron materials for sintering.

Table 11 shows a thermal profile and sinter made by sintering using 100% sinter grade ore, while coke breeze was used for sintering. Due to lack of fuel (3% of coke in mixture), there were low temperatures of 600–900°C in the sintered layer. The sinter had unacceptable properties – only some microgranules were connected. The

Iron burden	Coke breeze (%)	Photo	Mechanism	Temperatures (°C)
Iron ore (100%)	3		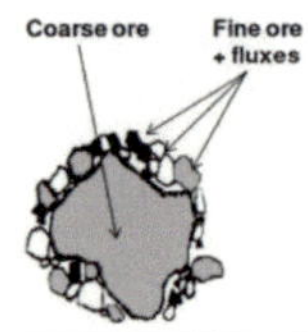	600–900
Iron ore (100%)	5		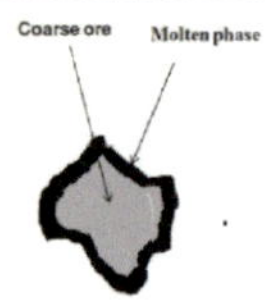	1150–1320
Iron ore (100%)	7		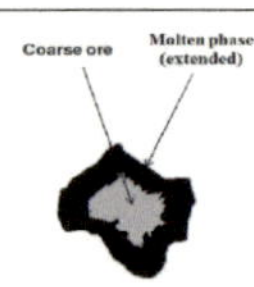	1180–1420
Iron concentrate (100%)	4		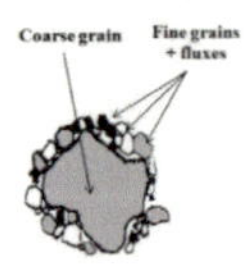	530–1020
Iron concentrate (100%)	6		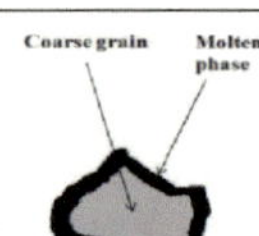	1130–1350
Iron concentrate (100%)	8		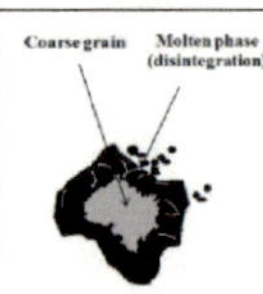	1210–1440

Table 11.
Production of sinters in laboratory conditions.

essence of the sintering process – i.e. producing the sinter with the required production, qualitative and quantitative parameters, was not accomplished in this case. The theoretical minimal quantity of fuel is always required for sinter production, which depends on the physical–chemical properties of input iron bearing raw materials and basicity of the produced sinter. In the presence of 5% of coke in sintering mix, standard temperatures of 1150–1320°C were achieved in the sintered layer with 100% of sinter grade ore, resulting in sinter with the required properties. On the surface of individual grains, the melt was formed, and due to the low viscosity of the liquid phase, multiple grains were bound together producing a sinter. At 7% of coke in mixture, high temperatures (up to about 1420°C) were reached in the sintered layer, resulting in sinter with the extended melting phase. Similar connections were found in the sintering of iron concentrates, with the fact that at higher fuel content (8% of coke) the sinter was disintegrated. A high FeO content was determined in this sinter (14.23%), which was a significant increase compared to FeO content in the charge (5.2%). Since the added fuel was probably not uniformly distributed into the individual grains (some were already partially sintered), there might have been microvolumes with a higher proportion of fuel. In these volumes, reducing conditions were created with high temperatures (up to about 1440°C), under which higher Fe_2O_3 and Fe_3O_4 oxides were reduced to FeO. In some samples of the sinter, break-up of sinter was observed. The disintegration of the sinter is sometimes associated with the formation of dicalcium silicate, sometimes with higher content of fuel. When using poor iron concentrate Nižná Slaná in the mixtures, higher amounts of fuel had to be used and the properties of the sinter were not adequate for the blast furnace process.

The achieved results from laboratory experiments can be summarized in the following points:

- Theory and balance calculations of the sintering process defined fuel (coke breeze) were the main source of gaseous emissions of CO and CO_2 in the flue gases.

- Due to the production of a highly basic sinter, a certain amount of CO_2 passes into the flue gas also through the dissociation of carbonates present in the burden (approx. 3–7% depending on the basicity).

- By reducing the underpressure, the temperature level in the sintered layer increases, the yield of the produced sinter is increased and the required quality parameters of the produced sinter are achieved.

- It has been found that increasing the fuel in the sintering charge increases the FeO content in the sinter. This fact is also confirmed by the effect of increasing FeO in the sinter with an increase in temperatures in the sintered layer.

- By increasing the ratio of concentrate/ore in the sintering burden (from 1.2 to 1.8) there was a decrease in the yield of the produced sinter.

- Technological recommendations were proposed in terms of achieving the required qualitative and quantitative parameters of the sintering process with emphasis on reducing CO and CO_2 emissions. Among the most important can be mentioned regular control of the permeability of the sintering charge, control and regulation of the vertical sintering rate in operating conditions, optimization of the underpressure, etc.) [10].

1.3 The use of traditional and alternative carbonaceous fuels in the production of sinters

1.3.1 Replacement of coke breeze with biomass

The ways of using biomass for energy purposes are predominantly predetermined by its physical and chemical properties. The biomass is characterized by the relatively high and frequently variable water content, which significantly affects the energy properties of biomass fuel. The various types of biomass (primarily of wood and plant origin) have much higher content of volatile matter than coke breeze. Analyses of the various types of biomass that have been used in the production of iron-bearing sinter in Slovakia shown in **Table 12** [10–13].

Carbonaceous materials	Moisture [%]	Ash [%]	Volatile matter [%]	Fixed carbon [%]	Sulfur [%]	Calorific value [MJ/kg]
Coke breeze	1.5	14.5	3.5	82.0	0.59	28.16
Charcoal	4.9	3.5	8.2	88.3	0.05	30.46
Walnut shells	9.6	0.7	81.3	18.0	0.05	16.90
Oak sawdust	7.1	1.5	83.4	15.1	0.05	16.56
Pine sawdust	13.6	0.9	85.6	13.5	0.05	15.94
Lignin	8.6	3.4	67.90	20.1	0.17	23.14

Table 12.
Analysis of selected carbonaceous materials which have been used in sintering in the Slovakia [10–13].

Type of fuel	CS [%]	Photograph of sinter	Sintering zone	Sinter characteristic	Ecological aspects
Coke breeze	0			Standard qualitative parameters	Standard parameters
Charcoal	20			Standard qualitative parameters	Lower emissions of CO_x, NO_x
Charcoal	44	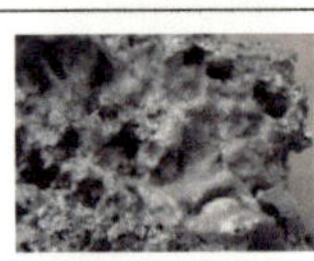		Higher volume of melt, excellent strength	Lower emissions of CO_x, NO_x, SO_2
Charcoal	50	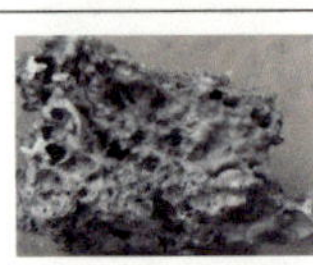		Higher inhomogeneity, standard qualitative parameters	Lower emissions of CO_x, NO_x, SO_2
Charcoal	86			Low volume of melt, worse qualitative parameters	Lower emissions of CO_x, NO_x

Type of fuel	CS [%]	Photograph of sinter	Sintering zone	Sinter characteristic	Ecological aspects
Oak sawdust	20			Standard qualitative parameters	Lower emissions of CO_x, SO_2
Oak sawdust	44			Low volume of melt, unacceptable qualitative parameters	Lower emissions of SO_2
Pine sawdust	8			Standard qualitative parameters	Lower emissions of CO_x, SO_2
Pine sawdust	20			Low volume of melt, unacceptable qualitative parameters	Lower emissions of CO_x, SO_2
Nut shells	8			Standard qualitative parameters	Lower emissions of NO_x, SO_2
Nut shells	20			Standard qualitative parameters	Lower emissions of NO_x, SO_2
Nut shells	50			Higher inhomogeneity, standard qualitative parameters	Lower emissions of CO_x, NO_x, SO_2
Lignin	20			Standard qualitative parameters	Lower emissions of NO_x, SO_2
Lignin	50			Low volume of melt, worse qualitative parameters	lower emissions of CO_x, NO_x, SO_2
Lignin	86			Low volume of melt, unacceptable qualitative parameters	Lower emissions of CO_x, SO_2

CS = coke substitution.

Table 13.
Characteristic of sintering of iron materials with biomass.

Sinter			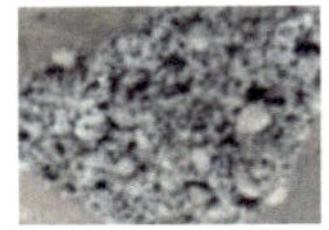				
		Coke powder (100%)	**Charcoal (86%)**	**Nut shells (20%)**	**Lignin (20%)**	**Oak sawdust (20%)**	**Pine sawdust (20%)**
Analysis (wt.%)	Fe	50.94	51.05	50.85	53.43	49.53	50.94
	FeO	8.35	5.32	6.76	7.55	6.33	5.08
	Fe_2O_3	66.71	63.22	61.07	67.54	63.79	67.24
	Mn	0.07	0.05	0.08	0.06	0.07	0.06
	SiO_2	9.88	10.64	11.65	7.13	10.27	11.04
	Al_2O_3	0.77	0.72	0.79	1.05	0.68	0.69
	CaO	10.29	10.64	12.51	12.54	11.83	12.41
	MgO	2.04	2.75	2.16	1.85	2.66	2.54
	P	0.021	0.026	0.020	0.061	0.024	0.042
	S	0.024	0.028	0.021	0.022	0.011	0.020
	K_2O	0.048	0.064	0.050	0.045	0.032	0.043
	C	0.14	0.23	0.21	0.22	0.14	0.18
Grain (mm)	> 25	40.17	44.01	42.36	47.42	24.38	25.17
	25–10	22.88	15.45	25.55	19.43	22.60	16.49
	10–5	14.63	14.28	14.27	11.70	20.38	18.53
	< 5	22.32	26.26	17.82	21.45	32.64	39.80
	d_A	16.71	16.53	17.64	17.85	13.00	12.19
Tumbler index (%)	+ 6.3 mm	62	62	58	65	54	50
Abrasion index (%)	- 0.5 mm	7	10	9	8	10	11

Sinter			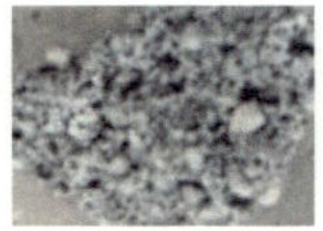	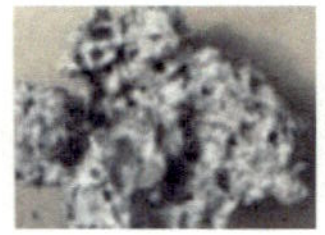		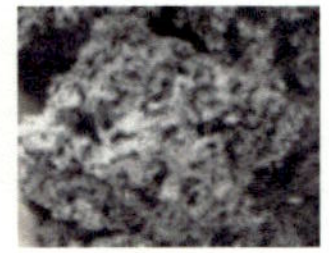	
		Coke powder (100%)	**Charcoal (86%)**	**Nut shells (20%)**	**Lignin (20%)**	**Oak sawdust (20%)**	**Pine sawdust (20%)**
Production yield (%)	P	78	76	80	79	70	60
Reducibility (%/min)	dR/dt	1.15	1.19	0.94	1.05	1.16	1.15
Mineralogy[*] (%)	XRD	Fe ox = 60 Si = 14 CaFe = 13 NA = 5	Fe ox = 51 Si = 14 CaFe = 24 NA = 9	Fe ox = 54 Si = 20 CaFe = 21 NA = 6	Fe ox = 57 Si = 16 CaFe = 19 NA = 7	Fe ox = 57 Si = 16 CaFe = 19 NA = 7	Fe ox = 57 Si = 16 CaFe = 19 NA = 7

[*]*Fe ox – iron oxides, Si – silicates, CaFe – calcium ferrites, NA – non-assimilated phases.*

Table 14.
Properties of sinters with coke substitution.

1.3.2 The influence of carbonaceous fuels on the quality of the sinter and on the economic and ecological parameters of the sintering of iron ores and concentrates

Starting from an analysis of the considered biofuels (**Table 12**) and proposed methodology for the performance of experiments, coke breeze was partly replaced with a defined quantity of individual biomass types. The materials whose composition is given in the **Tables 5, 10** were used for the experiments. **Tables 13, 14** give characteristics of sinters produced with biomass substitution of coke breeze. It can be seen in some experiments with biomass standard quality parameters on several sinters and lower emissions of CO_x, NO_x and SO_2 are achieved. It is possible to substitute about 10–20% of coke breeze by individual types of biomass in the sintering process. Combustion of plant biomass mainly depends on the carbon structure of the cellulose, hemicellulose and lignin, which differ from the amorphous carbon in coke (coke breeze). The maximum temperatures in the sintering process are lower with biomass than with actual coke breeze [11]. Biomass fuels can burn more quickly than coke breeze due to their high porosity and large interface area, while there is a significant increase in the vertical speed of sintering. The combustion of biomass decreases the maximum temperature and abbreviates the holding time at the high temperature. Lower temperatures in the sintered layer observed with the addition of biomass can also be attributed to the condensation of volatile organic compounds [11]. These compounds can eventually be converted into a phase similar to ash and reduce the heat transfer in the direction of burning.

The content of Fe_{TOT} in laboratory prepared sinters with biomass does not change considerably compared to the standard (reference – with 100% coke breeze) sinter and is within the interval of about 51–53%. The phase composition (mineralogy) of selected sinters is qualitatively comparable, while the differences are observable in the quantity of individual phases, as shown in **Table 14**. A sinter without biomass contains a higher proportion of iron oxides and sinters with biomass have more silicates and calcium ferrites. Compared to standard sinter, the increase in the share of calcium ferrites can be noticed in selected sinters with biomass. The microstructure of the standard sinter mainly consists of primary magnetite and hematite. Forms of silicoferrites of calcium and aluminum – SFCA are also visible in the microstructure to a small extent. In the microstructure of the sinter with biomass, there is also a visible area of the unsintered surface, which has been identified as lime. Forms of calcium silicates are also visible in the microstructure. The sinter with the substitution of coke breeze by charcoal reached the highest value of reducibility estimated using dR/dt ratio, as seen from **Table 14**. The sinter with the substitution of coke breeze by nutshells has the lowest value of reducibility. Sinters with coke breeze substituted by charcoal and sawdust have a similar reducibility to the reference sinter without any fuel substitution. Higher reducibility relates to sample properties such as its low content of FeO and its high porosity [14], which is related with sinters with the substitution of coke breeze by charcoal and sawdust.

2. Conclusions

In this chapter the properties of iron ores and concentrates were specified. These properties are very important for their efficient processing in the process of sinter and pig iron production. It is important to comprehensively evaluate these raw materials. The evaluation of the properties of iron ores shows that the best ores for blast furnace process have a suitable particle size distribution (10–40 mm), good chemical and mineralogical composition (especially hematite and magnetite, minimum of carbonate and silicate phases), they are well reducible (R_{i60} below 100 min.) and stable after temperature tests (+6.3 mm above 70%). Sinter grade ores and concentrates are characterized as iron ore raw materials with the required granulometry and chemical and mineralogical composition. In general, the richness of concentrates is in the range of 65–70%, while sinter grade ores have this interval wider and shifted slightly lower (55–67%). The larger grains of sinter grade ore are practically free of impurities and have a relatively homogeneous structure. In addition to iron oxides, the smaller sinter grade ore grains also contain impurities in the form of silicon and aluminum oxides. Undesirable impurities are mainly sulfur, phosphorus, zinc, lead, arsenic, copper, sodium, potassium, which are chemically bound in the minerals of the iron ores.

A new direction in the sintering process is showing the replacement of coke with biomass. It is a more environmentally friendly way of production and practice will show that it will also be more economical in the future. It can be seen in some experiments with biomass standard quality parameters on several sinters and lower emissions of CO_x, NO_x and SO_2 are achieved. It is possible to substitute about 10–20% of coke breeze by individual types of biomass in the sintering process. Properties of input iron bearing and basic raw materials are indeed crucial factors that affect the final quality of sinter, but no less important are the properties and the amount of carbonaceous fuel and high-temperature sintering technology. Progress in the production of pig iron in blast furnaces can be achieved by improving the quality of burden, especially iron ores and sinters. A significant improvement in the performance of blast furnaces is currently achieved by using a burden with a modified chemical and particle size distribution, which allows a more complete use of the chemical and thermal energy of the reducing gas.

Acknowledgements

This research was funded by [APVV] Slovak Research and Development Agency, Slovak Republic number APVV–16-0513.

Conflict of interest

The authors declare no conflict of interest.

Author details

Jaroslav Legemza*, Róbert Findorák, Mária Fröhlichová and Martina Džupková
Institute of Metallurgy, Faculty of Materials, Metallurgy and Recycling, Technical University of Košice, Slovakia

*Address all correspondence to: jaroslav.legemza@tuke.sk

References

[1] https://www.usgs.gov/centers/nmic/iron-ore-statistics-and-information

[2] Majerčák, Š.; Majerčáková, A.: Monograph, Vysokopecná vsádzka, ALFA Bratislava, Slovakia, 1986, p. 273

[3] Vitek, V.: Katalóg železných kusových rúd a pelet, Internal document, VÚHŽ Dobrá, Czech republik,1982–1998, p. 87

[4] Findorák, R.: Dissertation thesis, Metodika stanovenia vhodnosti vsádzkových surovín pre výrobu aglomerátu v požadovanej kvalite s minimálnymi nákladmi, Košice, 2009, p. 132

[5] Kret, J.: Vliv alkálií na výrobu surového železa ve vysoké peci. Hutnické listy, 2000, No.4, p. 10–14. ISSN 0018–8069

[6] Król, L.: Konstrukcja i urzandzenia wielkiego pieca. Katowice, Wydawnictwo Politechniki Ślaskiej, 1989, p. 481 ISBN 83–216–0809-4

[7] Buzek, J. a kol.: Alkalia w procesie wielkopiecowym. In: Acta Metallurgica Slovaca, 1999, No.5, p.55–60

[8] Dawson, P.: Recent developments in iron ore sintering and sinter quality, Part 2 Research studies on sintering and sinter quality. In: Ironmaking and Steelmaking. 1994, No.2, p. 137–143

[9] Fröhlichová,M., Ivanišin, D., Mašlejová, A., Findorák, R., Legemza, J.: Iron-Ore Sintering Process Optimization, Archives of Metallurgy and Materials 60(4), 2015, p. 2895–2899

[10] Legemza, J.; Fröhlichová, M.; Findorák, R. Biomass and Carbon Fuels in Metallurgy; CRC Press: Boca Raton, FL, USA, 2019. P. 292

[11] Legemza, J, Frőhlichová, M.; Findorák, R.; Džupková, M.; Modelling of Mass and Thermal Balance and Simulation of Iron Sintering Process with Biomass, Metals, 2019, 9, p. 1010–1028

[12] Findorák R; Frőhlichová M.; Legemza J.; Monograph, Biomasa v aglomeračnom procese, TUKE, Slovakia, 2020, p.127

[13] Frőhlichová, M.; Findorák, R.; Legemza, J.; Džupková, M. The Fusion characteristics of Ashes from Lignin and the Coke breeze, Arch. Metall. Mater., 2018, No.3, p. 1523–1530

[14] Jursova, S.; Pustejovska, P.; Brozova, S. Study on reducibility and porosity of metallurgical sinter, Alexandria Engineering Journal, 2018, No. 57, p. 1657–1664

Chapter 5

Application of X-Ray Diffraction to Study Mineralogical Dependence of Reduction: Disintegration Indices RDI of Blast Furnace Sinters

Hanna Krztoń and Janusz Stecko

Abstract

The aim of this research was to continue an examination of influence of mineral components of blast furnace sinters on their quality. Two of reduction-disintegration indices RDI were taken into account: static resistance to degradation RDI-$1_{+6.3}$ and static susceptibility to degradation RDI-$1_{-3.15}$. X-ray diffraction was used for phase identification and the Rietveld method was applied to study quantitative dependence. Static susceptibility to degradation RDI-$1_{-3.15}$ showed clearly dependence on quantitative mineral composition, namely on quantities of magnetite, silicates and slag phases. Static resistance to degradation RDI-$1_{+6.3}$ was also dependent on fractions of magnetite and silicates.

Keywords: Rietveld method, blast furnace sinters, mineralogy, reduction-disintegration indices RDI

1. Introduction

The history of integrated factory's steel product usually begins from a preparation of a sinter being one of the main component of a blast-furnace charge. In blast-furnace practice a sinter is a product after high temperature treatment of a mixture of iron ores and their concentrates, fluxing agents (dolomite, lime) and coke breeze. Sintering process takes place in a range of 1200–1300°C and causes partial melting of individual grains of all sinter's components, forming a compact product. The temperature has to be lower than a melting point of the mixture. There are some phase transformations and/or degradation of minerals which accompany the sintering processes. As a result, a mineralogy of a sinter is different than a mineralogy of starting components.

Generally, the mineralogy of sinters can be thought as a mineralogy of two types of compounds: oxides and silicates, including also silicate glass. The presence and the amount of individual minerals depend on the properties of and the conditions of technology (practice). One of the most important factor, characterizing a sinter, is its basicity; the value of this parameter is defined as a ratio of a content of calcium (given as calcium oxide) to a content of silicon (given as silicon dioxide).

The basicity of a sinter is the main reason of diversity in observed mineral composition and – consequently – in properties of the sinter.

The identification of mineral components can be done using various experimental techniques from which the X-ray diffraction seems to be the most important. Every crystalline chemical compound (mineral) produces its own, typical X-ray diffraction pattern which can be used in identification of it in a multiphase mixture. The individual patterns are kept in Powder Diffraction File, updated every year by the International Centre for Diffraction Data and having now (2020) more than 400,000 data of inorganic compounds. In many cases, the data for one particular compound contains also information about crystallography of this compound. It means that a calculation of a theoretical pattern of this compound is possible. In X-ray diffraction method, one can assume that the observed pattern of a multiphase powder mixture is a sum of weighed patterns of individual compounds (minerals). This assumption gives the possibility to apply the Rietveld method to study the multiphase mixtures and to calculate the fractions of individual components of the mixture.

The Rietveld method was presented for the first time in 1967 by H. Rietveld to refine crystallographic structure of a crystalline compound, using neutron diffraction [1, 2]. R. A. Young and D. B. Wiles applied the method to X-ray diffraction in 1981 [3, 4] and the next application to quantitative phase analysis was introduced by Hill and Howard in 1987 [5] and Bish and Howard in 1988 [6]. The crucial advantage of the Rietveld method in studying and quantifying mixtures is the ability to analyze the overlapped reflections. Overlapping is the most important problem in mixtures, especially in the case of low symmetry of constituents. The number of reflections can reach some hundreds in a typical 2Theta range of measurements. There is also another problem – a contribution of intensities of many small reflections to an observed intensity of a background of an experimental X-ray diffraction pattern and consequently some difficulties in refining the background. This is also the case of the presence of amorphous component. This matter was solved in the Siroquant software [7] in which the shape and intensity of the background is not refined but manually removed from a pattern. Owing to the above, the Rietveld method gives the unique opportunity for a precise quantitative description of blast furnace sinters.

The quality of sinter's pieces can be described using different indices which have to be determined by some technical tests and experiments, according to international standards [8].

The tests are carried out in conditions which simulate blast-furnace's conditions in upper part of its shaft. A sample of a sinter of 500 g is placed into a furnace heated up to 500÷550°C and is exposed to a reducing gas for 60 minutes. Then the sinter is cooled in an atmosphere of inert gas and is subjected to tumbling. The disintegration sinter's pieces gives grains of various sizes which are screened to three groups. The following indices, according to International Standard ISO 4696-1:2007, are determined:

Static resistance to degradation (the ratio of reduced sinter with size larger than 6.3 mm after tumbling test to reduced sinter)

$$RDI - 1_{+6.3} = \frac{m_1}{m_0} \times 100\% \tag{1}$$

Static susceptibility to degradation (the difference between reduced sinter and the sum of reduced sinter with size larger than 6.3 mm and reduced sinter with size larger than 3.15 mm divided by reduced sinter)

$$RDI-1_{-3.15}=\frac{m_0-(m_1+m_2)}{m_0}\times 100\% \tag{2}$$

Static grindability (the difference between reduced sinter and the sum of reduced sinter with size larger than 6.3 mm, reduced sinter with size larger than 3.15 mm and reduced sinter with size larger than 0.5 mm divided by reduced sinter)

$$RDI-1_{-0.5}=\frac{m_0-(m_1+m_2+m_3)}{m_0}\times 100\% \tag{3}$$

where: m_0 [g] – mass of a sample after reduction before tumbling m_1 [g] – mass of oversize particles remained on a screen of 6.3 mm after tumbling m_2 [g] – mass of oversize particles remained on a screen of 3.15 mm after tumbling.

m_3 [g] – mass of oversize particles remained on a screen 0.5 mm after tumbling.

From technological point of view, the most important index is a static susceptibility to degradation and a question of dependence of its value on a mineralogical composition of a sinter is still to be answered. Previous examinations showed that there was a connection between quantities of mineral components of sinters and their reducibilities [9] and this work is a continuation of studying a dependence of quantitative mineral composition of blast furnace sinters on their quality.

1.1 Materials

All sinters were prepared from raw materials which were used in Polish steel plants. There were two kinds of iron ores and three kinds of concentrates of other iron ores (**Figure 1**). Six different mixtures were prepared, each one consisted of different combinations of ores and concentrates but with a planned basicity (**Table 1**). Basicity was calculated as a ratio of calcium content recalculated to CaO to silicon content given as SiO_2. Sintering process was carried out in laboratory conditions with application of lime and dolomite in a conventional sinter pot of a diameter of 490 mm [10]. Ten laboratory tests were done for each kind of mixture (**Figure 2**); then three samples characterized by optimal moisture with the highest productivity were selected for further investigations (**Table 2**). The RDI values were calculated according to the procedure given above.

Figure 1.
Samples of hematite lump ore (a), magnetite concentrate (b) and sinter (c).

	Series A	Series B	Series C	Series D	Series E	Series F
Basicity CaO/SiO2	1,20	1,20	1,50	1,50	0,90	1,20

Table 1.
The planned basicity values of each kind of mixtures.

Figure 2.
The experimental site for sintering iron ores and waste products in Sieć Badawcza Łukasiewicz - Instytut Metalurgii Żelaza.

Sinter RDI values	A1	A2	A3	B1	B2	B3	C1	C2	C3
RDI-1+6.3	39,5	40,2	45,9	46,2	50,2	51,5	44,3	51,1	55,8
RDI-1-3.15	29,9	28,7	26,1	26	22,6	22,4	21,3	17,2	18,8
Sinter RDI values	**D1**	**D2**	**D3**	**E1**	**E2**	**E3**	**F1**	**F2**	**F3**
RDI-1+6.3	54,9	55,2	52,8	39,7	36,6	39	51,3	53,7	52,6
RDI-1-3.15	14,3	15,2	14,9	32,4	32,9	33,5	20,8	19	19,1

Table 2.
The RDI values of the chosen sinters.

2. Methods

In XRD method, the first step was to reduce a crystallite size of the sinters, using Fritsch Puverisette 0 mill and also a mortar and a pestle. The measurements of diffraction patterns were carried out using a PANalytical Empyrean diffractometer with a PIXcel solid state detector and filtered Co K_a radiation (Fe filter on a diffracted beam). The range of 2θ was 10°- 100° with step size Δ2θ = 0.02626° and time/step 800 s. The identification of phases was done according to International Centre for Diffraction Data (ICDD) Powder Diffraction File PDF-4+. The quantities of individual phases were calculated by means of the Rietveld method and Siroquant software [7]. An amorphous component was detected and quantified after adding a small addition of corundum (certified Standard Reference Material SRM No. 676a from National Institute of Standards and Technology NIST, USA) to an initial sample of a sinter, next a homogenization and a repeated measurement.

3. Results

Generally, two groups of minerals can be distinguished in sinters: oxides and silicates. The examined sinters contained two iron oxides: hematite Fe_2O_3 and magnetite Fe_3O_4 and one silicon dioxide – quartz SiO_2. Only traces of wuestite FeO were found. The group of silicates consisted of calcium silicates (larnite β-$CaSiO_4$, γ-$CaSiO_4$, wollastonite $CaSiO_3$), calcium-iron silicates (hedenbergite $Ca_{0.5}Fe_{1.5}Si_2O_6$, kirschsteinite $CaFeSiO_4$), magnesium silicate $MgSiO_3$ and so called silica glass (amorphous component). There were also three so called slag phases identified in the sinters: $Ca_{2.3}Mg_{0.8}Al_{1.5}Fe_{8.3}Si_{1.1}O_{20}$ (SFCA phase), $Ca_{3.18}Al_{1.34}Fe_{15.48}O_{28}$ (SFCA-1 phase) and $Ca_{2.45}Fe_{9.2}Al_{1.74}Si_{0.6}O_{20}$ (SFCA Mg – free phase).

All measured diffraction patterns of the examined sinters were characterized by a strong complexity because of many reflexes of individual phases and overlapping of the reflexes. In **Figure 2**, a part of a diffraction pattern of one of the sinters is shown and the positions of reflexes of all crystalline components, according to PDF-4+ standard data, are clearly visible. In a typical experimental range of 2θ (10° – 100°2θ) there were about:

- 90 reflexes of larnite β-Ca_2SiO_4
- 105 reflexes of SFCA $Ca_{2.3}Mg_{0.8}Al_{1.5}Fe_{8.3}Si_{1.1}O_{20}$
- 56 reflexes of SFCA-1 $Ca_{3.18}Al_{1.34}Fe_{15.48}O_{28}$
- 77 reflexes of hedenbergite $Ca_{0.5}Fe_{1.5}Si_2O_6$

and also reflexes of other constituents. This was a result of low symmetries of some of the phases forming a sinter. There are all identified phases and their space groups in **Table 3**.

With such a complexity of the diffraction patterns, the only solution was to apply the Rietveld method [1–6]. A model of a sinter file was created, containing all structural data of all identified crystalline components. Two numerical criteria of fit were taken into account: χ^2 and *R-pattern*. The global parameter of the refinement was 2θ- *zero.* Pearson VII was chosen as an analytical profile function. For each phase, the refined parameters were: scale, lattice parameters, full-width-at-half-maximum (FWHM) parameters. The background was removed manually.

The example of the Rietveld refinement is presented in **Figure 3**. The upper line is the experimental pattern (as measured). The higher intensity of the background in the range of lower angles is indicated and it is concerned with a presence of amorphous component. Below, there is the experimental pattern after removing background (points) and the calculated refined diffraction pattern (solid line). At the bottom, a difference plot is shown.

During the refinements, the difference plots were carefully observed. In some cases, the identification of a phase of a small content was possible during the analysis of a shape of a difference plot.

Phase	Space group	Phase	Space group
Magnetite	Fd-3 m	Hedenbergite	C2/c
Hematite	R-3c	Kirschsteinite	Pmnb
Larnite	P21/n	MgSiO3	Pbcn
γ-CaSiO4	Pnma	SFCA phase	P-1
Wollastonite	P-1	SFCA-1 phase	P-1
Wuestite	Fm-3 m	Quartz	P3221

Table 3.
Space groups of sinters' crystalline components.

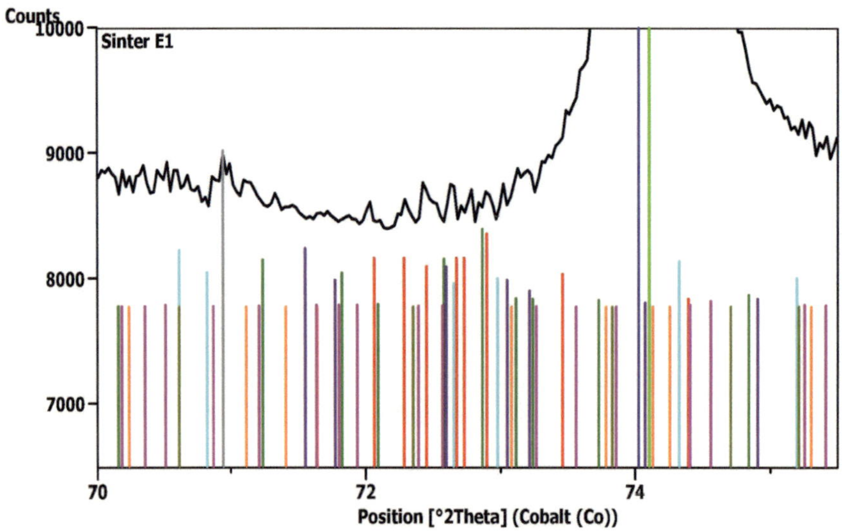

Figure 3.
An example of the Rietveld refinement of a diffraction pattern of one of a sinter. The upper line is the experimental pattern (as measured). Below, there is the experimental pattern after removing background (points) and the calculated refined diffraction pattern (solid line). The higher intensity of the background in the range of lower angles is concerned with a presence of amorphous component. At the bottom, a difference plot is shown.

4. Discussion

No significant differences were observed in the qualitative mineral composition of the sinters. A diversity in some minor components (silicates) was determined, excluding dicalcium silicates which were present in all sinters. The chemical composition and qualitative mineralogical investigations could not provide enough information about any dependence between a mineral composition and the quality indices of sinters. The next step was to apply quantitative phase analysis to check whether – or not – a quantitative dependence existed. The application of the Rietveld method to quantitative phase analysis in sinters was the only possible solution to obtain sensible results. The one of the most important problems in the refinements was the removing of background. The shape of background in sinters is not linear, the presence of amorphous component can influence on both intensity and shape of the background. Moreover, the intensity and shapes of reflexes of minor components (as it can be seen clearly in **Figure 4**, where the positions of

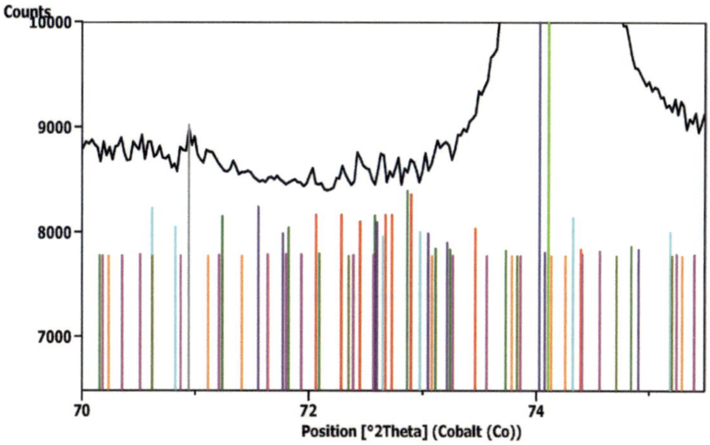

Figure 4.
A small part of a diffraction pattern of one of the sinters. The marks show the positions of reflexes of all components of the sinter, according to their standard data taken from PDF-4+ file. Complexity of the pattern is caused by the number of reflexes.

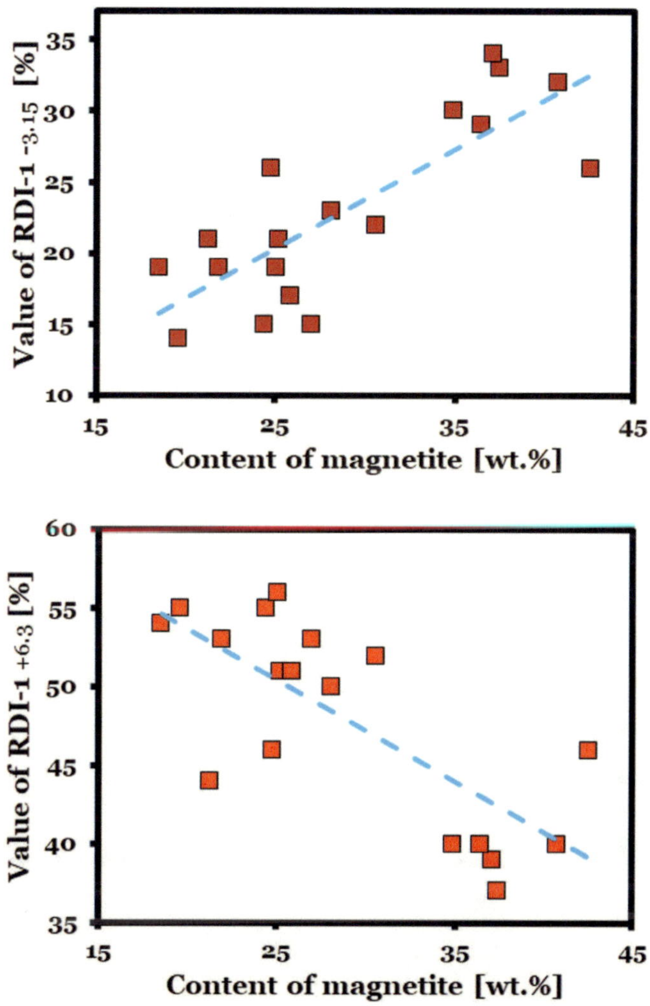

Figure 5.
The RDI-1$_{-3.15}$ (top) and RDI-1$_{+6.3}$ (bottom) dependence on contents of magnetite in the examined sinters. The growing tendency of RDI-1$_{-3.15}$ with growing content of magnetite and the opposite dependence of RDI-1$_{+6.3}$ are shown.

reflexes of all crystalline components are shown in a small range of a diffraction pattern of one of the sinters) also take part in forming the background. In this case, the careful manual removing of background before start of refinements should be used instead of the refinement of background.

The quantitative results show a noticeable dependence of RDI-$1_{-3.15}$ on contents of all kinds of minerals, excluding amorphous component (**Figures 5–9**). The reverse dependence was observed for main components, namely magnetite and hematite.

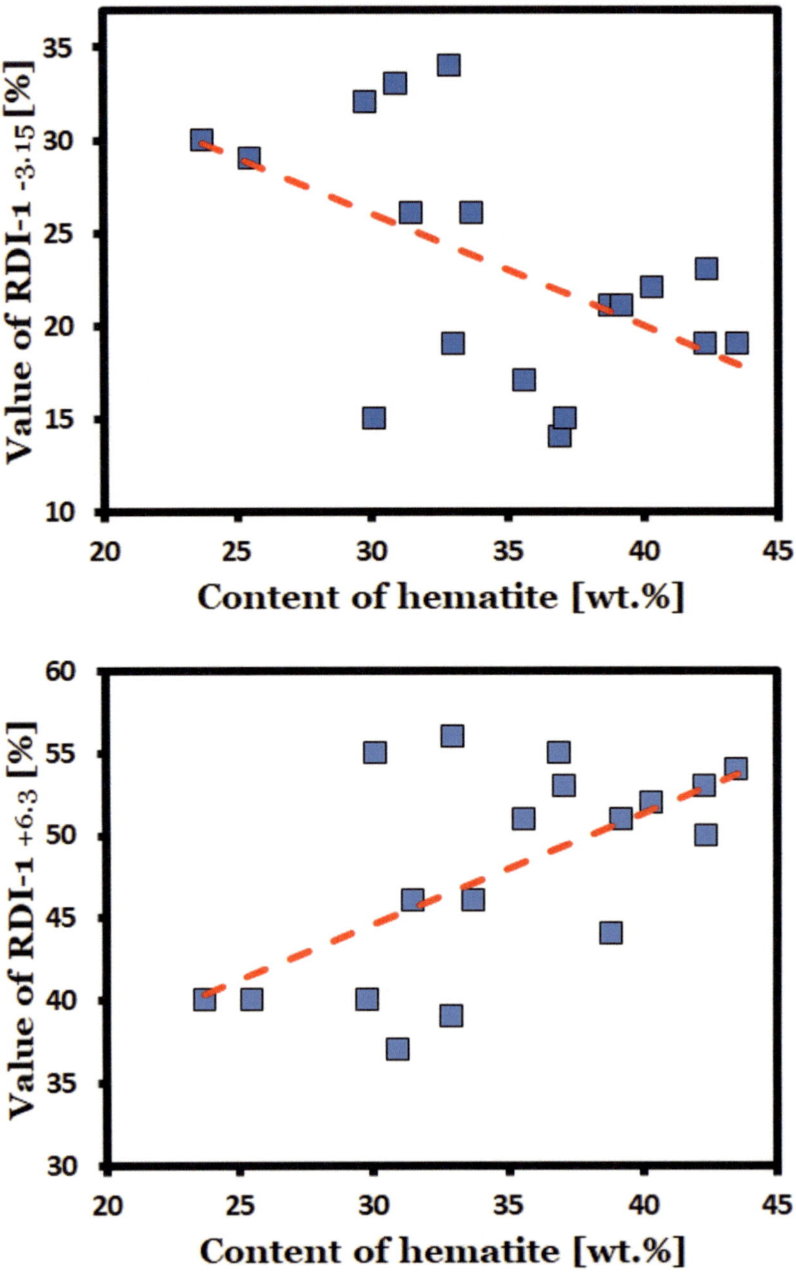

Figure 6.
*The RDI-$1_{-3.15}$ (top) and RDI-$1_{+6.3}$ (bottom) dependence on contents of hematite in the examined sinters. The lowering tendency of RDI-$1_{-3.15}$ with growing content of hematite (in a range of values of RDI-$1_{-3.15}$ from 20 to 30) and the opposite dependence of RDI-$1_{+6.3}$ are shown in **Figure 7**.*

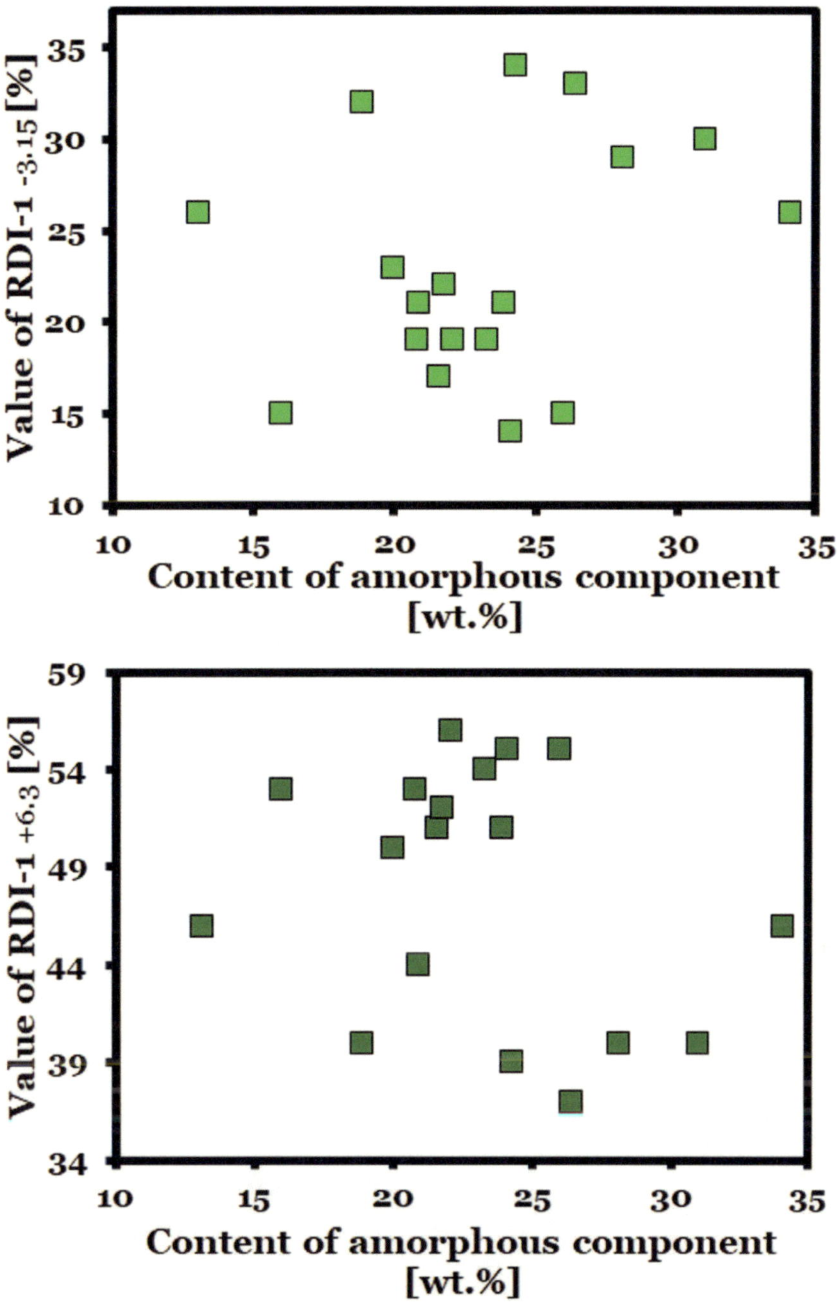

Figure 7.
The RDI-1$_{-3.15}$ (top) and RDI-1$_{+6.3}$ (bottom) dependence on contents of amorphous component in the examined sinters. No tendency is observed for both indices.

The lower value of RDI-1$_{-3.15}$ (the lower means the better) corresponded with higher values of content of hematite (but not in a full range of values of RDI-1$_{-3.15}$) and with lower values of magnetite. No clear tendency was noticed for amorphous component. Unexpectedly the quite high content of dicalcium silicates and also slag phases accompanied the lowest values of RDI-1$_{-3.15}$. As an explanation of the results, one could assume that both, silicates and slags, could form a kind of binder among grains of other minerals and because of it had an effect on obtaining the low values of RDI-1$_{-3.15}$.

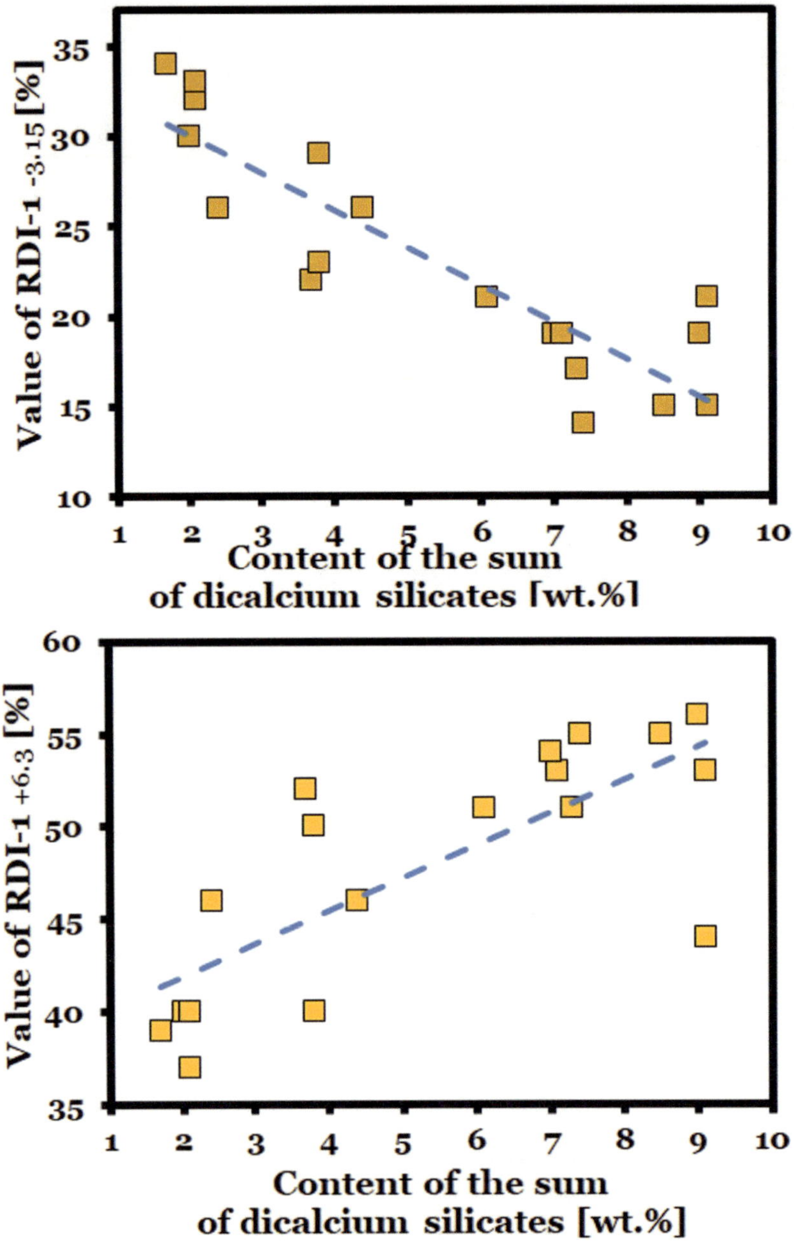

Figure 8.
The RDI-$1_{-3.15}$ (top) and RDI-$1_{+6.3}$ (bottom) dependence on contents of a sum of dicalcium silicates in the examined sinters. The linear, lowering tendency with increasing content of dicalcium silicates is seen for RDI-$1_{-3.15}$ No tendency is seen for RDI-$1_{+6.3}$.

The dependence of RDI-$1_{+6.3}$ values on fractions of mineral components of sinters was also considered (**Figures 5–9**). As in of RDI-$1_{-3.15}$ case, magnetite and hematite contents showed a linear dependence from RDI-$1_{+6.3}$ values; the increasing values of RDI-$1_{+6.3}$ were related with lower values of magnetite contents and higher values of hematite fractions. The other minerals did not show so clear regularities.

The observed dependencies of RDI-$1_{-3.15}$ on basicity is shown in **Figure 10** - the higher value of basicity the lower RDI-$1_{-3.15}$ was obtained what can be thought as

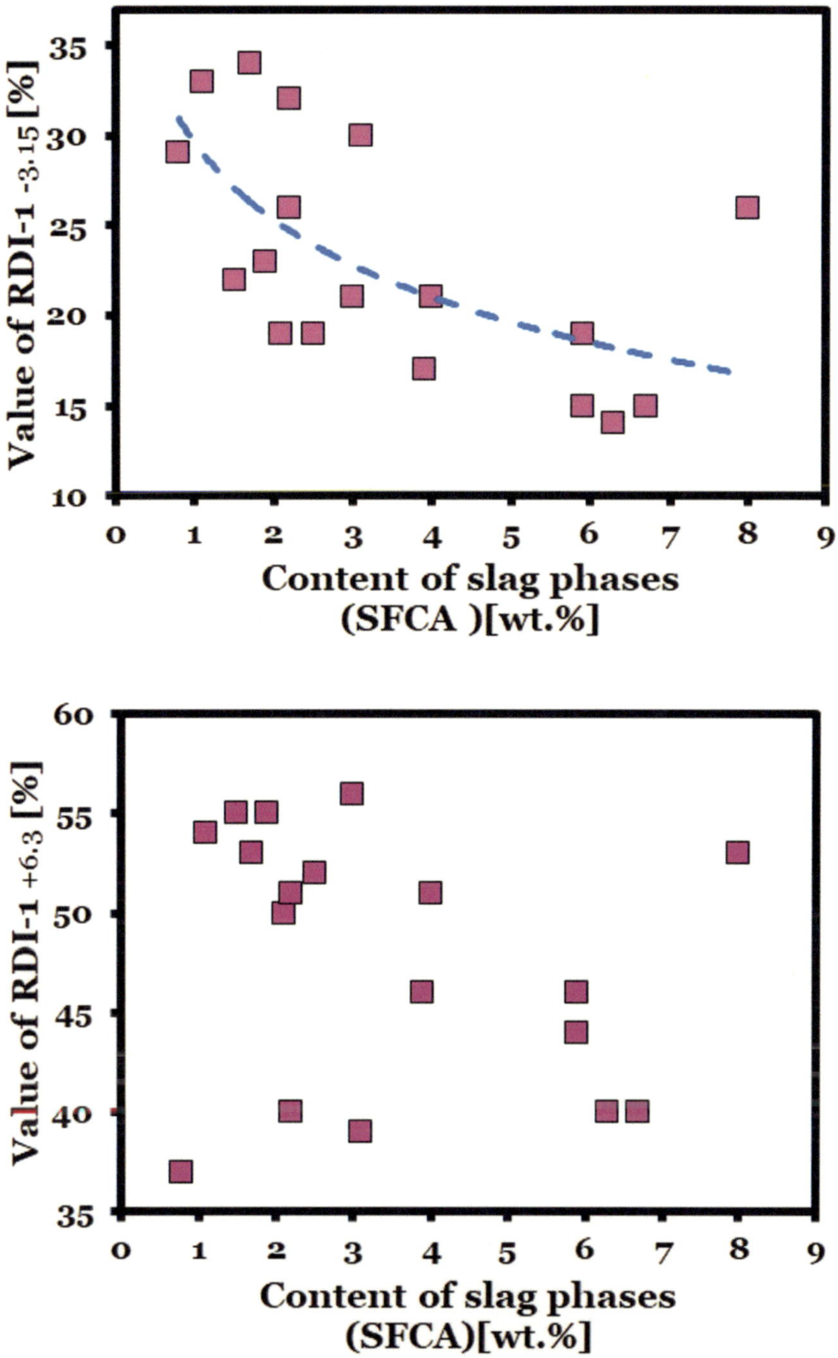

Figure 9.
The RDI-1$_{-3.15}$ (top) and RDI-1$_{+6.3}$ (bottom) dependence on contents of a sum of slag phases in the examined sinters. It can be seen that higher content of slag phases lowers the value of RDI-1$_{-3.15}$.

a beneficial result. In comparison, the rise of FeO content (not as a mineral, but as Fe^{2+} recalculated to oxide) gives higher values of RDI-1$_{-3.15}$ (**Figure 10**) what is an unfavorable tendency. This result is in accordance with the results for magnetite ($FeO.Fe_2O_3$) (**Figure 5**) which is the main source of Fe^{2+} in sinters

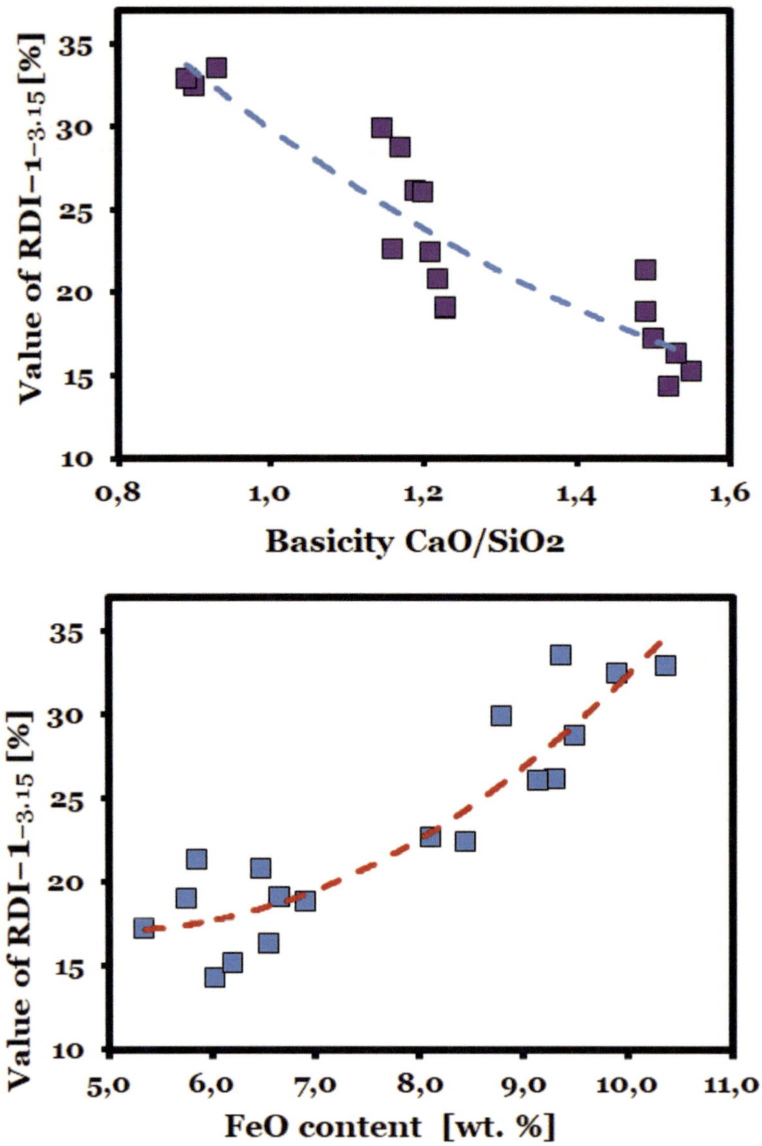

Figure 10.
The dependence of RDI-1$_{-3.15}$ *on basicity* CaO/SiO_2 *(top) and on FeO content (bottom) in sinters with calculated tendency lines.*

5. Conclusions

- The identification procedure showed no significant differences in qualitative phase composition of the examined sinters. There were only small discrepancies in occurrence of some silicate components (as e.g. magnesium silicate), but their contents were about 1 wt.%, so their influence on properties of the sinters was not crucial.

- There is a clear dependence between static susceptibility to degradation RDI-1$_{-3.15}$ and quantities of mineral components of the sinters. This is not observed for all minerals - amorphous component do not show any kind of correlation with values of RDI-1$_{-3.15}$. The correlation of the content of hematite and corresponding values of RDI-$1_{-3.15}$ is not evident – it can be seen only in a range from 20 to 30 of values of RDI-$1_{-3.15}$. and in this range is decreasing with growing contents of hematite.

- The growing dependence of values of RDI-$1_{-3.15}$ with higher contents of magnetite is easily seen and it is in a good agreement with the results of studies of influence of FeO content (Fe^{2+} recalculated to oxide) on values of RDI-$1_{-3.15}$ as the main source of Fe^{2+} in sinters is magnetite.

- The reverse tendency can be observed for dicalcium silicates and also for slag phases. The increasing content of these minerals is accompanied with lowering values of RDI-$1_{-3.15}$. It can be correlated with basicity values – the Ca^{2+} atoms are mostly present in the above mentioned type of minerals.

Acknowledgements

This work was supported by Polish Ministry of Science and Higher Education according to contract no. 3889/E-139/S/2017.

Author details

Hanna Krztoń* and Janusz Stecko
Sieć Badawcza Łukasiewicz - Instytut Metalurgii Żelaza, Gliwice, Poland

*Address all correspondence to: hanna.krzton@imz.pl

References

[1] Rietveld, H.M., A Profile Refinement Method for Nuclear and Magnetic Structures, J. Appl. Crystallogr. 2, 1969. p. 65-71.

[2] Rietveld, H.M., Line profiles of neutron powder-diffraction peaks for structure refinement, Acta Crystallog. **22**, 1967. p. 151-152.

[3] Wiles D. B., Young R. A., A new computer program for Rietveld analysis of X-ray powder diffraction patterns, J. Appl. Crystallogr., **14**, 1981. p. 149-151

[4] R. A. Young, Introduction to the Rietveld method, in: R. A. Young (ed.), The Rietveld Method, Oxford University Press 1993

[5] Hill R. J., Howard C. J., Quantitative phase analysis from neutron powder diffraction data using the Rietveld method, J. Appl. Crystallogr., **20**, 1987. p. 467-474

[6] Bish D.L., Howard S. A. Quantitative phase analysis using the Rietveld method, J. Appl. Crystallogr. **21**, 1988. p. 86-91

[7] SIROQUANT™, Quantitative XRD software, ver. 3.0 for Windows 2007

[8] ISO 4696-1:2007 Iron ores for blast furnace feedstocks -- Determination of low-temperature reduction-disintegration indices by static method -- Part 1: Reduction with CO, CO_2, H_2 and N_2

[9] Krztoń H., Stecko J., Kania Z., The quantitative dependence of reducibility on mineralogical composition in iron ore sinters (blast furnace sinters), Acta Physica Polonica A, **130** (4), 2016. p. 1147-1150

[10] PN-ISO 8263:1999 Iron ore fines- Method for presentation of the results of sintering tests. Warszawa The Polish Committee for Standardization 1999

Chapter 6

Plasma Processing of Iron Ore

Sumant Kumar Samal, Manoja Kumar Mohanty, Subash Chandra Mishra and Bhagiratha Mishra

Abstract

The depletion of high-grade ore minerals and the scarcity of fossil fuel reserves are challenging factors for metallurgical industries in the future. Also, extensive mining for increased steel demand results in the generation of fines often found unsuitable for use as direct feedstock for the production of metals and alloys. Apart from mines waste, the other major sources of fine minerals are leftover in charge burdens, sludges, and dust generated in the high-temperature process. Sludge and fines generated during beneficiation of ore add to this woe, as the outcomes of beneficiation plants for lean ores show better yield for fine particles. The utilization of lean ore and wastes in iron making requires wide research and adopting new advanced technologies for quality production with time-saving operations. The application of thermal plasma in mineral processing has several advantages that can overcome the current industrial metal extraction barriers. The present study demonstrates the thermal plasma for the processing of different iron-bearing minerals and its feasibility for metal extraction.

Keywords: thermal Plasma, blue dust, siliceous iron ore, manganiferous iron ore, recovery

1. Introduction

The global production of crude steel exceeds 1869 million tonnes in the year 2019. Steelmaking viz. carbon steel and alloy steel is a multistep process where iron ore is the starting material used for iron making. Blast furnace iron making is mostly adopted by industries throughout the globe [1, 2]. Production of DRI for smelting in EAF is an alternative for iron ore reduction. There are several problems that persist with the economy of iron and ferroalloys production, and it depends upon three major factors viz. material, process, and product. The characteristics of ore minerals decide the process kinetics, and hence product quality and yield. There are several problems that still persist, as the following needs to be resolved.

i. Ore minerals: The quality of iron ore plays a significant role as the cost of raw ore attributes about 40% of the total production cost. The mined ore needs to be in the specified size range for individual furnace types, which is accomplished by crushing and sizing. The crushing and washing of bulk ore generate a substantial amount of fines viz. micro and macro fines, which cannot be fed directly into a furnace as it affects the porosity of charge burden. Moreover, it increases process cost comprising of agglomeration and heat treatment before extraction.

ii. Gangue content: The excavated ore always includes gangue contents viz. alumina, silica, and magnesia along with alkali, sulfur, and phosphorus. The type and quantity of gangue affect the entire process kinetics in terms of metallic yield and quality. The primary ore needs to be upgraded through various separation techniques, i.e., physical, gravitational, etc. which is critical for high gangue amounts.

iii. Mineral phases: The mineral phases present in the ore are of interest as the entire extraction process is dependent on the various minerals present in the parent ore. Minerals in the ore are detected as metal oxides, hydroxides, carbonates, and also associated with gangue as silicates, aluminates. The silicates and aluminates phases are not only difficult to reduce but also consume high flux and energy for which ores with high content of such phases are commonly discarded at the mines site itself. The decomposition of hydroxides and carbonates results in higher coke consumption. The presence of alkali not only affects the process but also has a high impact on refractory linings.

iv. Process: With the increased demand for steel across the globe in the scenario of unaffordability of high grade ores, research on the applicability of fines, dust, and other industry by-products has become essential in order to control the depletion of earth minerals. These fines are agglomerated through pelletization, sintering, and briquetting routes, which has various drawbacks in terms of production rate, energy consumption, charging, and environmental impact. Ore and agglomerate must be of suitable for minimal transportation loss, high-temperature sustainability, and low disintegration rate. The porosity, density, and crushing strength of agglomerate must be adequate in order to achieve a higher reduction rate and metallic yield. If such properties are not in the predesigned range, it can cost higher and affect smooth operation by promoting fines generation and hinders Boudouard reaction.

The utilization of lean ore and wastes in iron making requires wide research and adopting new advanced technologies for quality production with time-saving operations.

2. Plasma

It is not unusual to refer plasma as the fourth state of matter as it is an ionized gas comprised of molecules, atoms, ions (in their ground or in various excited states), electrons, and photons. Plasma possesses a unique property known as quasi-neutrality since plasma is electrically neutral. In contrast to an ordinary gas, plasma encloses free electric charges that are commonly produced from the gas itself by a variety of ionization processes. In a steady-state situation, the rate of ionization in the plasma is balanced by the rate of recombination. Depending upon the energy content of the plasma, the degree of ionization may be so high that virtually no neutral particles are left, i.e., the plasma becomes fully ionized [3, 4].

2.1 Classification

Since plasma is a broad topic as concerned, all together plasmas are classified into three main categories [5]:

- CTE plasmas (complete thermodynamic equilibrium)
- LTE plasmas (local thermodynamic equilibrium)
- Non-LTE plasmas (nonlocal thermodynamic equilibrium)

Among the above three types, CTE plasmas are used for thermonuclear fusion experiments. The latter two types are used as laboratory plasmas and also implemented for industrial purposes like MINTEK, South Africa. Again according to density and energy, typical plasmas are categorized as shown in **Figure 1**.

Plasmas generated by electron and photon belong to the nonlocal thermodynamic equilibrium category. LTE plasmas are also called as hot plasmas or thermal plasmas and non-LTE plasmas as cold plasmas or non-thermal plasmas. Based on temperature, plasmas are subcategorized into two groups, i.e., low-temperature plasma and high-temperature plasma. Plasmas with temperatures below 10^{5}°K or in other words, energies less than 10 eV per particle are to be called as low-temperature plasmas. Beyond this limit, it is said to be high-temperature plasma. It is also not unusual for plasma to be called as per its gas name, i.e., oxygen plasma, argon plasma, nitrogen plasma or argon-nitrogen plasma, etc.

2.2 Plasma chemistry

Plasma chemistry refers to the thermodynamic characteristics of various plasma forming gases. Both monoatomic and diatomic gases like argon, helium, neon, nitrogen, oxygen, hydrogen, carbon monoxide, carbon dioxide, air, and a mixture of gases are used as plasma forming gases. The relation between energy and temperature of some commonly used monoatomic and diatomic gases are shown in **Figure 2**.

The diatomic molecules require 90 to 200 kcal mole-1 to dissociate between 4000 to 10,000°K, while ionization requires 340 to 600 kcal mole-1 between 10,000 to 30,000°K [5]. The upper practical limit of flame temperature is about

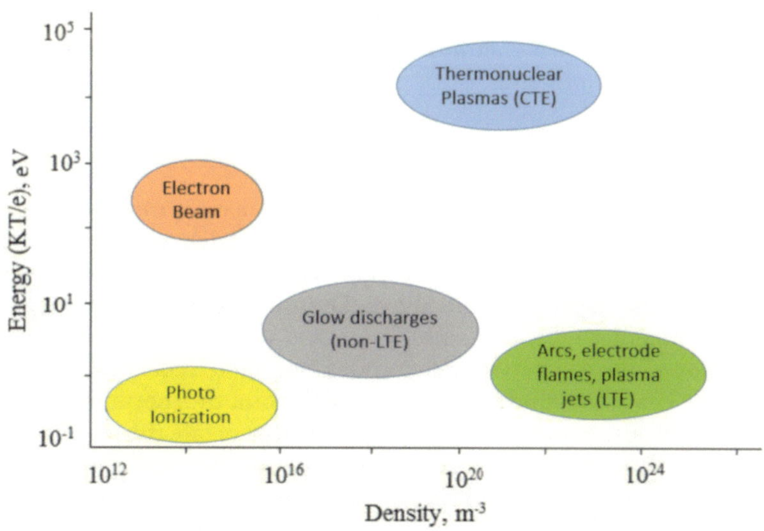

Figure 1.
Typical plasmas characterized by their energies and densities.

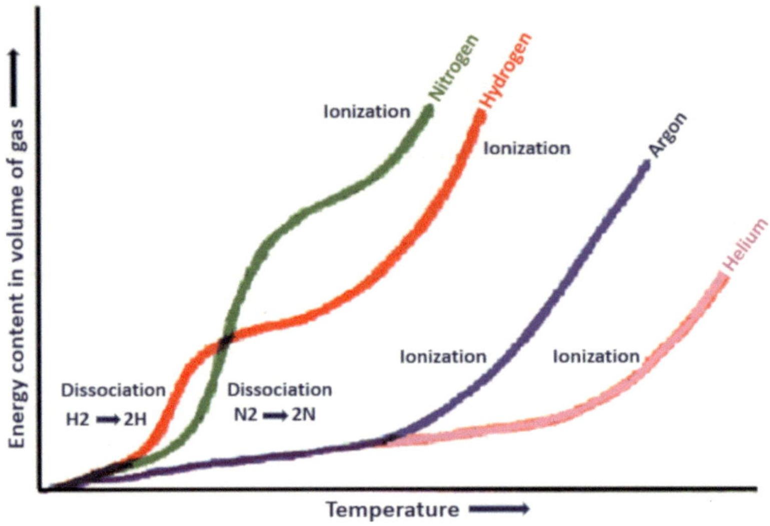

Figure 2.
Temperature and energy relationship of various plasma gases.

3500°K, where molecules begin to dissociate, while the lower limit of plasma temperatures is about 10,000°K. As most laboratory plasmas are heated electrically, their temperatures will lie in the bottom end of the ionization curve, i.e., above 10,000°K for diatomic gases. For any process operating below 1000°K, an air-fuel flame (~2000°K) or an oxygen-fuel flame (~3000°K) will have a high percentage of energy available for the process. However, for the reaction occurring at 2500°K, only one-sixth of energy contained in an oxygen flame will be available, and rest must be either wasted or recovered in the expensive heat exchangers. On the other hand, a plasma flame composed of atomic nitrogen at 10,000°K would have more than 90% of its energy available above 2500°K. This high energy efficiency may more than offset the economic advantage that combustion energy over electrical energy; certainly, this advantage will increase as electrical energy becomes cheaper while fossil energy gets more expensive. Although by utilizing plasma high temperature can be achieved with the liberation of huge heat energy in a chemical reaction, plasma gases are generally not used as reactants in the reaction.

2.3 Generation of plasma

Thermal arc plasmas are generated by striking an electric arc between two or more electrodes. They are characterized by high current densities (greater than 100 A/cm^2) and are more luminous than other types of discharges, especially when operated at atmospheric pressure and above. Thermal arcs can be initiated in several ways. Two common methods are electrode contact, which produces a short circuit, or pre-ionization of the gap between electrodes by a high-frequency spark. The cathode must be heated beyond 3500 °K, at which point the thermionic emission of electrons begins, generating the charge carriers that create the plasma state [3]. Cold cathodes are cylindrical and made of heavily cooled copper, iron, or copper alloy while high-temperature cathodes are usually rod-shaped and made of thorium, tungsten, or graphite. Thermal arc plasma torches can operate in two modes, i.e., non-transferred and transferred arc. If the plasma torch having two electrodes designed in such a way that hot gas emerges through one electrode and then heated by the flame is called non-transferred. If there is only one electrode in the torch and

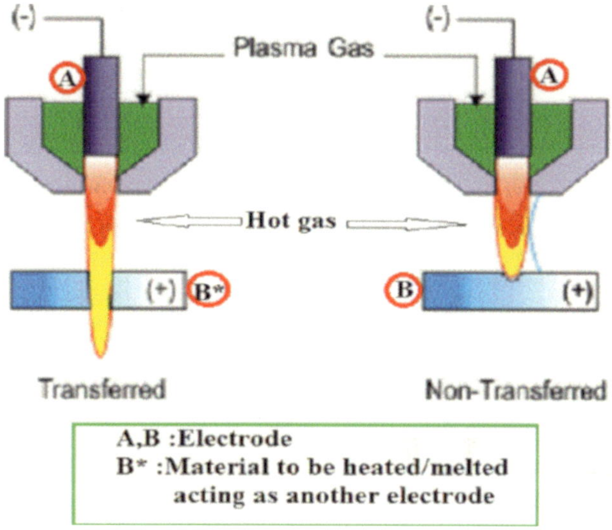

Figure 3.
Schematic diagram of transferred and non-transferred plasma torches.

material to be heated/melted acts as another electrode, then it is said to be transferred. The schematic of both transferred and non-transferred arc plasma torches are shown in **Figure 3**.

2.4 Application of plasma

In the last two decades, plasma has claimed to be an emerging solution to numerous processes due to its unique features and hence implemented in various sectors [4, 6–10]. Plasma finds significant industrial applications viz. melting, smelting, smelting and reduction, remelting and refining, spark plasma sintering, surface modification, and surface coating.

2.5 Advantages of plasma over conventional processes

Although there are a lot of many advantageous aspects behind the utilization of plasma, some of the important features are given in **Table 1**.

2.6 Application in iron making

Many researchers investigated the applicability of plasma in iron and steel making [9, 11, 12]. In general, plasma is used as a heat source instead of reductant itself, as the percentage of the degree of reduction lags behind when utilized as a reductant. The selection of the type of plasma and preferred operating parameters, along with the type of reductant, is a crucial factor that needs to be considered sensibly in relation to the treating of material. The wrong choice can affect both troubleshooting and also processing costs.

Criteria for selection must be based on answering many questions, which comprises;

i. Type of reducing agent (carboneous or any other)

ii. Type of plasma forming gas (inert, self-reducing, self-burning or helps in burning)

High efficiency	Since a huge amount of energy in the form of heat is available by utilization of plasma, high throughput can be achieved.
Long-range of melting materials	Since high temperature can be achieved in a reaction by using plasma, almost all materials can be melted in this process. Although its commercial use to melt and process metals is well known, the method is less known as a method of melting glass.
Feed capability	This process is independent of the size, shape, and composition of feed material.
Transient process	Due to the release of huge heat energy that a particular reaction requires at a specific temperature, plasma stands ahead of any other process to respond to the changes in a shorter period.
High energy fluxes	Higher temperatures with extreme jet velocities and greater thermal conductivities of plasma gases are the key factors that result in high energy fluxes. Smaller furnace dimensions with high smelting capacity are a unique aspect of using plasma.
Independent energy source	The flexibility of control over feed rate and power independently and input power is not limited by the electrical conductivity of feed material to be melted or smelted. Hence greater freedom of choice with respect to charge composition is available by using plasma.
Gas flow control	Unlike combustion systems, the gas flow rate, temperature, and energy input are not interdependent, and gas flow rate and temperature can be controlled separately irrespective of energy input.
Gas environment control	Energy can be provided to the system with desired oxygen potential to ensure oxidizing, reducing, or inert gas conditions independently without taking temperature into account.
Electrical energy-intensive process	Minimization of the usage of fossil fuel energy and conservation of fossil fuel can be made.
High energy transfer to slag layer	Plasma jet is directed towards slag layer and significantly increases the metallization rate.
Purity level in product	The purity level of the final product through plasma processing is very high.

Table 1.
Advantageous aspects of thermal plasma.

iii. Type of process (melting, smelting or smelting reduction)

iv. Process duration

v. Process environment (open-air, inert or vacuum)

vi. Feed rate

vii. Power control

3. Description of the plasma furnace

The schematic diagram of 30 kW DC extended arc plasma reactor used for this study is shown in **Figure 4** [13].

On top of the reactor, the plasma torch is attached in the downward direction. The plasma torch contains a hollow cylindrical graphite crucible with 145 mm outer diameter, wall thickness 15 mm, and 300 mm high that serves as the anode. A hollow graphite rod of 400 mm long and 5 mm inner and 35 mm outer diameter serves as the cathode. The graphite rod end is tapered to a conical shape for superior

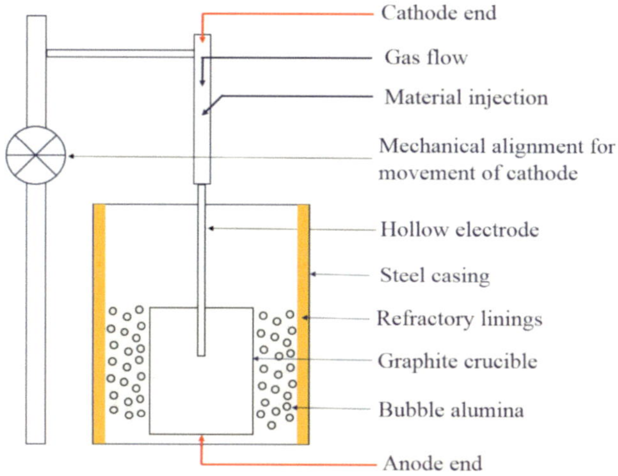

Figure 4.
Schematic diagram of DC extended arc plasma reactor.

electron emission. The hollow structure of the cathode has been designed to have provisions for gas flow. The material to be processed was placed in the anode crucible bed, and the arc was initiated by shorting the cathode and the crucible bottom wall (graphite plate). The arc length was increased by raising the cathode rod up suitably within the crucible to heat the charge placed in the crucible. The power supply and power control unit is designed to vary the necessary voltage and current, enabling easy and smooth control of experimental conditions. Voltage and current can be altered over a range of 0–100 V and 0–500 A, respectively. The gas supply unit facilitates plasma forming gases, i.e., oxygen, argon, nitrogen, methane, coke oven gas. Besides, the mixture of above gases can be utilized as plasma forming gas. Gas flow control consisting of digital indicators helps in not only measuring gas flow rate but also governing a suitable flow of gases as per experiment performed and stands as a key parameter. The gas flow rate can be varied from 0 to 15 LPM. Heat insulating materials are placed in between the steel casting and reaction chamber.

Several prerequisite steps have to be done before feeding samples into the reaction chamber. Initially, the crucible was cleaned in order to avoid any other material contained in the crucible to be reacted with samples. The hollow tapered graphite rod was fitted in such a way that it points towards the center of the reaction chamber. After checking no leakage in the crucible, it was placed in the space provided in steel casting. Bubble alumina was poured in spacing between the reaction chamber and reaction chamber that acts as a heat-insulating medium. The power supply was then provided, and proper arcing between cathode and anode was tested. The gas supply is then connected to the cathode passage, and plasma forming gas was purged into the reaction chamber for 1 minute to displace atmospheric air. After that, the power supply and plasma forming gas supply both supplied simultaneously, and the required voltage and current maintained. Then sample feed was poured into the hot reaction chamber as per our requirement.

4. Plasma processing of iron-bearing minerals

The present study demonstrates the plasma processing of three iron-bearing minerals viz. blue dust, siliceous type iron ore, and manganiferous iron ore.

4.1 Plasma processing of blue dust

Blue dust is the purest form of iron oxide mineral (hematite) abundantly available in many states of India. For the present study, blue dust of Koira origin, Odisha, India, was collected, which is in fine form (150 μ). The chemical analysis of blue dust is given in **Table 2**. The ore is mainly composed of Fe_2O_3, and XRD analysis also confirmed the presence of single mineral hematite.

Plasma smelting operations were carried out for mixtures of blue dust and coke in argon and nitrogen ionizing atmosphere [14]. The coke percentage in charge mixture (500 gm) was varied from 5–20%. The plasma gas flow rate was maintained at 2.5 LPM.

The highest recovery rate exceeding 86% was achieved for using nitrogen as plasma forming gas. The recovery rates in argon plasma are comparatively less than those of nitrogen plasma. It is because of the diatomicity of N_2 gas, which liberates higher energy flux than the monoatomic gas Ar. The loss of Fe in the process involves loss accounted for in charging and splashing of metal droplets due to the high velocity of the plasma jet in the course of smelting. The loss of metal splashing is further minimized by adjustment of power input and controlling gas flow rate. The recovery rate attains 95% maximum in closed furnace type arrangements.

As the gangue in blue dust is low, the metallization (Fe) occurs in the absence of complex slag phases. Blue dust with different carbon percentages (i.e., 5, 10, 12, 15, and 20) smelted by using nitrogen plasma shows the change of ferrite, ferrite-cementite to fully pearlite structure, which can be attributed to the Hull-Mehl model of pearlitic transformation [15]. The silica in blue dust in the high reducing atmosphere reduces into SiO, observed in smelting tests as fumes. The smelting duration for the conversion of Fe_2O_3 into Fe was 17 minutes, which is several hours in BF iron making. Moreover, blast furnace limits the direct charging of blue dust to avoid lowering the porosity of charge burden, which increases process cost and affects smooth operation.

To use blue dust in BF, agglomeration and heat treatment are required. Although stiff vacuum extrusion briquetting avoids heat treatment, binder requirement is still essential. The cement and bentonite binder adds cost and also requires unnecessary slag generation and separation from the purest Fe_2O_3 ore.

The direct smelting of blue dust in thermal plasma has several advantages over conventional processes in terms of cost-saving operation, purity level in hot metal, and high production rate. The production cost will be much less for industrial large scale furnace and by using cheap gases such as methane, coke oven gas, etc.

4.2 Plasma processing of siliceous type iron ore

For this study, partially reduced briquettes made from iron minerals were collected from an industry in the vicinity of Rourkela, Odisha, India. Briquettes upon solid state reduction at 1250°C are partially melted which hinders further reduction at higher temperatures. The industrial trial of such briquettes in mini BF suggested its infeasible use for iron making due to high FeO loss in slag. The chemical composition of the briquette sample is given in **Table 3**.

The amount of silica and alumina in the briquette is about 16% in cumulative. XRD analysis detected wustite (FeO), fayalite (Fe_2SiO_4), and hercynite ($FeAl_2O_4$) as major phases in the briquette sample. The presence of such phases suggests the high affinity of FeO towards silica and alumina for which low melting fayalite forms, melts early and hinders CO gas passage to the core. Partial melting of briquette also affects the furnace operation and increases flux addition, and hence increases the process cost.

Constituents	In Wt. %
Fe_2O_3	96.87
SiO_2	0.45
Al_2O_3	0.21
MgO	Trace
LOI	1.48

Table 2.
Chemical composition of blue dust.

Constituents	In Wt. %
Fe_T	72.2
SiO_2	8.6
Al_2O_3	7.2
MgO	0.63
CaO	1.4
TiO_2	0.4
Others	9.57

Table 3.
Chemical composition of briquette.

Here, an attempt was made for the utilization of these briquettes for the value addition with maximized extraction [16]. Since plasma processing does not restrict the slag chemistry, briquettes were smelted with and without flux (CaO). Initial trials with flux addition targeting melilite slag (CaO-MgO-Al_2O_3-SiO_2) improved Fe recovery in metal. For the CaO/SiO_2 ratio in the range of 0.9–1.0, metallic yield exceeds 88%. The flow characteristics of such slag allow a better reduction in the slag layer where unreduced Fe-oxides are more promptly metalized.

Another approach was aimed at the direct smelting of briquettes without adjustment of slag chemistry. Since the briquettes are composed of fayalite, additional coke was provided for the reduction of silicon along with iron. These briquettes were smelted for a longer period than previous slag practice. The metallic recovery was appreciably higher, i.e., exceeds 94% by using nitrogen as plasma forming gas.

Phase and microstructure evolution confirms the formation of the iron silicide (Fe_3Si) phase in the alloy along with Fe. These ferrosilicon alloys can be used for deoxidation purposes, which is of greater value than metallic Fe.

This study suggests that the utilization of silicate-based iron minerals are more suitable for ferrosilicon production rather than iron making. Although the energy consumption is a little higher for FeSi production from these briquettes, flux consumption and melting of excess slag can be eliminated. Moreover, the product (FeSi) cost puts importance on its feasible production.

4.3 Plasma processing of manganiferous iron ore

Manganiferous iron ore is the type of lean manganese ore containing a maximum about 10–15% of Mn. These are of less importance in ferromanganese production; however, reduction roasting and magnetic separation improve Mn/Fe ratio. The primary objective of such a process is to reduce Fe_2O_3 into Fe_3O_4, which easily separates as magnetic particles. However, the feasibility of the upgrading process becomes questionable when both iron and manganese oxides are in associated form, i.e., bixbyite $(Fe, Mn)O_3$ mineral.

As an alternative, these ores are subjected to smelting for obtaining FeMn alloy with low Mn content. It is a cost-saving operation, and smelting operations can be carried out even in BF. The complexity arises for such ores with high gangue amount, which affects the extraction kinetics by forming silicates, aluminates, and/or complex mixtures phases.

In the present study, lean manganese ore was collected from Joda valley, Odisha, India. The ore is in fine form and is being discarded as waste at the mines site itself. The initial assessment of the ore through wet chemical analysis indicated that the ore contains about 17% of alumina and 9% of silica. The Mn content in the fines is about 12%, which falls into the manganiferous category. The reduction studies of such briquettes evidenced the formation of hercynite, galaxite, fayalite-manganon, and spessartine phases at different temperatures. These phases lower the reducibility of the ore and also deteriorate the physical and mechanical properties of the agglomerate.

Here an attempt was made to utilize these fines directly in thermal plasma, avoiding any agglomeration. Smelting of such ores by using other technologies results in poor Mn recovery (≈30%) and high FeO loss into slag; flux addition was essential.

The smelting of ore with flux addition targeting melilite and mayenite slags in ionizing atmosphere improved Mn recovery and was 80% maximum. Although plasma arc provides high energy flux, the slag chemistry also governed the process kinetics. By adjusting slag chemistry to a too basic slag lowered the activity of silica and alumna; however, the formation of high melting silicate compounds such as dicalcium silicate and tricalcium silicate increases the viscosity of the slag. The flowability of such slag hinders carbon contact with metal oxides and hence lowers the reducibility.

In the current scenario, ferromanganese production follows rich slag and discard slag practices. The rich slag retained in primary smelting (low fluxing) is further smelted in another step to produce silicomanganese or ferro-silicomanganese. In discard slag practices, the slag retained in primary smelting, which contains less than 15–30% MnO, is discarded.

The present study refers to the discard slag practice followed by plasma smelting with the highly basic slagging operation. As the ore contains high alumina, primary smelting similar to rich slag practice will result in slag with alumina bearing compounds, which will be difficult to reduce in the secondary smelting for obtaining silicomanganese. Moreover, the cost of smelting these high melting compounds will increase reductant, energy consumption, and also lower the furnace refractory life cycle.

The extraction of metals from these types of complex ores in single-stage smelting operation should be chosen in such a way that the slag can be used in secondary products such as cement.

5. Hydrogen plasma

In iron making, coke is used as a heat source and reductant. The application of plasma in iron making lowers CO_X emission for being used as plasma as a heat source. The reducibility of metal oxides by solid carbon or CO gas is lower than that of H_2. The use of methane as plasma forming gas is beneficial over argon or nitrogen from a cost perspective. However, ecofriendly gas emission in iron making is only possible through hydrogen plasma processing; the exit gas is water vapor, which reduces environmental pollution and will be much beneficial in impurity-free metal production [17, 18].

At present, research projects are being carried out for hydrogen reduction of iron ore in a pilot-scale, such as HYBRIT [19]. The primary installation of such reactors costs high; continuous improvements are essential. The primary beneficiation of iron ores will improve the purity of iron ore, which in turn will reduce the cost of the process.

6. Conclusion

The importance of plasma in iron making is discussed considering different types of ore minerals and its various aspects of processing. The freedom in size, composition, and smelting conditions required for complex ore minerals fits into the processing of iron ore in thermal plasma. The use of coke as a heat source in conventional iron making processes can be eliminated with the application of thermal plasma. The recovery rate and purity level in hot metal extracted from complex mines waste is noticeable higher by using thermal plasma. The future eco-friendly hydrogen plasma processing is of interest. Moreover, the use of hydrogen plasma can result in carbon-free metal/alloys, which can lower production costs by avoiding decarburization.

Author details

Sumant Kumar Samal[1*], Manoja Kumar Mohanty[1], Subash Chandra Mishra[1] and Bhagiratha Mishra[2]

1 National Institute of Technology, Rourkela, India

2 Suraj Products Limited, Rourkela, India

*Address all correspondence to: sumantnitr@gmail.com

References

[1] Seetharaman S, McLean A, Guthrie R, Sridhar S. Treatise on Process Metallurgy [Internet]. Vol. 1, Treatise on Process Metallurgy. Elsevier; 2013. 1-952 p. Available from: https://linkinghub.elsevier.com/retrieve/pii/C20100666910

[2] Chatterjee AG and A. Iron making And Steel making, Theory and Practice. first. PHI Learning Pvt. New Delhi: PHI Learning Private Limited; 2008. 161-167 p.

[3] Vladimir Dembovsky. Plasma Metallurgy: The Principles. Elsevier Science Ltd; 1985.

[4] Samal S. Thermal Plasma Processing of Ilmenite [Internet]. Cham: Springer International Publishing; 2018. (SpringerBriefs in Applied Sciences and Technology). Available from: http://link.springer.com/10.1007/978-3-319-70733-4

[5] Feinman J. Plasma Technology in Metallurgical Processing. Vol. 14, Iron and steel society. Iron and Steel Society; 1987. 33-36 p.

[6] Rykalin NN. Thermal plasma in extractive metallurgy. Pure and Applied Chemistry [Internet]. 1980 Jan 1;52(7):1801-1815. Available from: https://www.degruyter.com/view/journals/pac/52/7/article-p1801.xml

[7] Maske KU, Moore JJ. The application of plasmas to high temperature reduction metallurgy. High Temperature Technology [Internet]. 1982 Aug 6;1(1):51-63. Available from: https://www.tandfonline.com/doi/full/10.1080/02619180.1982.11753180

[8] Gauvin WH, Drouet MG, Munz RJ. Developments in Plasma Processes for Extractive Metallurgy. JOM [Internet]. 1987 Dec 26;39(12):14-17. Available from: http://link.springer.com/10.1007/BF03257563

[9] Jayasankar K, Ray PK, Chaubey AK, Padhi A, Satapathy BK, Mukherjee PS. Production of pig iron from red mud waste fines using thermal plasma technology. International Journal of Minerals, Metallurgy and Materials. 2012;19(8):679-684.

[10] J. Ronald Gonterman MAW. Plasma Melting Technology and applications. In: Wallenberger FT, Bingham PA, editors. Fiber glass and Glass Technology [Internet]. Boston, MA: Springer US; 2010. p. 431-51. Available from: http://link.springer.com/10.1007/978-1-4419-0736-3

[11] Badr K, Bäck E, Krieger W. Plasma Reduction of Iron Oxide by Methane Gas and its Process Up-scaling. steel research international [Internet]. 2007 Apr;78(4):275-280. Available from: http://doi.wiley.com/10.1002/srin.200705892

[12] Upadhya K, Moore JJ, Reid KJ. Application of Plasma Technology in Iron and Steelmaking. JOM: Journal of The Minerals, Metals & Materials Society. 1984;36(2):46-56.

[13] Samal SK, Mishra B, Mishra SC. Carboaluminothermic Production of Ferrotitanium from Ilmenite Through Thermal Plasma. Journal of sustainable metallurgy. 2020. DOI: 10.1007/s40831-020-00292-5

[14] Samal SK, Sindhoora LP, Mishra SC, Mishra B. Studies on plasma processing of blue dust. IOP Conference Series: Materials Science and Engineering [Internet]. 2015 Feb 19;75:012030. Available from: https://iopscience.iop.org/article/10.1088/1757-899X/75/1/012030

[15] Mehl RF. The Structure and Rate of Formation of Pearlite. Metallography, Microstructure, and Analysis. 2015;4(5):423-443.

[16] Swain B, Samal SK, Mohanty MK, Behera A, Mishra SC. Study of effective utilization of iron ore sinter through arc plasma. IOP Conference Series: Materials Science and Engineering [Internet]. 2018 Mar;338:012022. Available from: https://iopscience.iop.org/article/10.1088/1757-899X/338/1/012022

[17] Sabat KC, Murphy AB. Hydrogen Plasma Processing of Iron Ore. Metallurgical and Materials Transactions B: Process Metallurgy and Materials Processing Science. 2017;48(3):1561-1594.

[18] Rajput P, Sabat KC, Paramguru RK, Bhoi B, Mishra BK. Direct reduction of iron in low temperature hydrogen plasma. Ironmaking and Steelmaking. 2014;41(10):721-731.

[19] Åhman M, Olsson O, Vogl V, Nyqvist B, Maltais A, Nilsson LJ, et al. Hydrogen steelmaking for a low-carbon economy [internet]. 2018. Report No.: EESS report no 109. Available from: https://www.sei.org/wp-content/uploads/2018/09/hydrogen-steelmaking-for-a-low-carbon-economy.pdf

Chapter 7

Concentration and Microwave Radiated Reduction of Southeastern Anatolian Hematite and Limonite Ores—Reduced Iron Ore Production

Yildirim İsmail Tosun

Abstract

The concentration of low grade iron ore resources was evaluated by washing and reduction. The advanced concentration methods for low grade limonite and hematite iron ores of South Eastern Anatolian resources required such specific methods. The followed column flotation and magnetic separation, microwave radiated reduction of hematite slime and limonite sand orewere investigated on potential reducing treatment. The bubling fluidized bed allows more time to the heat radiation and conduction for reducing to the resistive ıron compounds. Furthermore, heavy limonite and iron oxide allowed sufficient intimate contact coal and biomass through surface pores in the bubbling fluidized bed furnace due to more pyrolysis gas desorption. Bubbling bath porosity decreased by temperature decrease. This research was included reduction in microwave of poor hematite and limonite ores in the microwave ovens, but through smaller tubing flows as sintering shaft plants following column flotation and scavangering operation. Two principle stages could still manage prospective pre reduction granule and pellet production in new sintering plants. There is a lack of energy side which one can produce reduced iron ore in advanced technology plants worldwide. However, for the low grade iron ores such as limonite and sideritic iron ores it was thought that microwave reduction technique was assumed that this could cut energy consumption in the metallurgy plants.

Keywords: microwave radiation, ıron waste slurries, reducing bath, bubling bath, concentration treatmen, sorbent bath, ferrite, waste ferrite, limonite slurries, microwave activation, reducing treatment, iron ore composts

1. Introduction

In 2017 1.63 billion tons of steel was produced in the world as given in **Table 1** and first share of that was China with 631 million ton first and even highest consumption with 736,8 million tons of pig iron and the other share countries were sequentially classified as Japan, USA, India and Russia. Iron ore production in the same years was also reached 2162 billion tons as given in **Table 2**. In the first ranks in iron ore production was first state Australia, Brazil, China, India and Russia was competed

Thousand tons	2015	2016	2017
Europian Union (28)	166,115	162,024	168,305
Turkey	31,517	33,163	37,524
Other Europian	35,778	37,601	42,203
CIS	101,552	102,108	100,933
North America	110,938	110,638	115,761
South America	43,899	40,220	43,693
Africa	13,701	13,099	15,053
Middle East	29,429	31,480	34,475
Asia	1,112,873	1,123,948	1,123,948
Oceania	5717	5837	5985
World	1,620,001	1,626,954	1,690,479

Table 1.
Crude steel production [world steel association, 2019].

Million tons	2015	2016	2017
Europian Union (28)	27,825	30,187	30,675
Turkey	9994	7520	6226
Other Europian	15,407	11,072	9526
CIS	195,298	189,064	177,906
North America	112,451	107,590	113,838
South America	448,122	465,811	465,590
Africa	93,760	90,559	85,657
Middle East	39,370	43,280	55,087
Asia	283,527	316,758	336,879
Oceania	814,403	861,521	887,365
World	2,030,164	2,115,842	2,162,524

Table 2.
Iron ore production [world steel association, 2019].

by even scrap trade. In these years in the world 88,664 million tons of reduced iron (**Table 3**), about 101 million tons scrap and sponge iron was imported [1, 2].

Iron ore reserves in our country distributed in Divriği, Bingöl, Kayseri regions. Besides these ores Kesikköprü, Balikesir and Adapazari different types of iron ore reserves are available. The low quality iron ore was extracted and mixed a certain weight rate amount with high quality ores evaluated in sinter making [3].

Iron ore is used in iron and steel plants. Turkish iron ore reserves can be absolute ore deposits 115 mill tons, apparant iron ore deposits 950 mill tons, potential iron ore deposits 432 millions determined in three groups as tons. Substantially expressing our country's iron ores as low quality, high titaniferrous iron ore is a serious mistake. Low grade iron ore deposits occurred in mainly Sivas, Malatya and Erzincan [4, 5].

In 2019, the recent years by Erdemir Plant of OYAK, in Turkey, produced pig iron in 4 iron blast furnaces used high grade (63% Fe total) imported iron ores mixed with at high rate over 5/1 local iron ores evaluated [4, 5]. Hence, this

Thousand tons	2015	2016	2017
Europian Union (28)	659	702	703
CIS	5436	5820	7200
North America	10,621	8644	9618
South America	2680	1691	1722
Africa	4025	4019	6152
Middle East	28,906	29,816	33,194
Asia	23,654	27,638	30,075
Oceania	5717	5837	5985
World	75,982	78,331	88,664

Table 3.
Direct reduced iron ore production [world steel association, 2019].

evaluation of low grade explored iron ore deposits prompted. The limonite ore deposites contained low grade Fe such as 25–42% Fe in South Eastern Anatolia needed enrichment for evaluation in blast furnaces after sizing suitable for use. Low grade hematite ore was concentrated to over 65% Fe grade by gravity and magnetic separation. The advanced methods were searched for low grade iron ore potential to evaluate. But some of the ore undesirable by integrated impurities and low iron content. It was partially operated in certain periods. An important part of these resources contains limonitic and sideritic iron ores and below 44% Fe as low grade hematite river sands. The limonite ore deposits range between 19 and 44% Fe. Low grade hematite ores 20% Fe grade could be evaluated in India and China. The limonite ores are enriched and pelletized [4, 5].

The pelleting plant projected in Divrigi contained a concentrator and pelleting facility established at a capacity of 1 million hematite ore. Regarding our iron ore qualities over undesirable contents such as Ti and Mn the resources was not projected in evaluation as long years since no study conducted on behalf of preparation cost and mining cost of iron ore.

India ranked first reduced iron ore use in world steel production with integrated iron and steelmaking waste gases in established in the direct reduction facility [5].

In Turkey, imported scrap at 9 million used for 22 million tons EAF iron and steel making instead of concentration and pelleting our low grade ores. Reduced ıron ore pellet plants in India [6–9]. produced 36.9 million tons in 2019 and varied total ore production of 232 million tons. Vertical gas shaft furnaces were commonly used in evaluation of reduced low grade hematite sands followed pelleting and imported low grade ores in this country directed to horizontal grate for production reduced sinters on grate-reduction using coal gas or steel making flue gas. Although reduced pellets were high grade 85% Fe total ore produced, sponge iron production was not commonjust 3 million tons.

The most effective and cost-effective technologies are needed for iron ore products in today's modern technologies. Turkish pig iron industry needs direct reduced iron ore products and related technologies and high quality at lower cost with various types of local iron resources should be investigated.

Şırnak and Hakkari region asphaltite coal combusted and ash discarded in the boiler bottom contains 15% limonite at fine sand size below 2 mm from furnace [10]. The rivers and some local land soil contained below 22% Fe grade hematite at slime size of below 150 micron simply excavated as river sand and exported to Iran. In this study this low quality hematite and limonite ores were evaluated for steelmaking reduced iron ore [11–13]. The gravity concentration was affective in limonite and hematite sands. The Humphrey spirals were also used in the washing of Şırnak asphaltite coal bottom ash. The limonite sands gave high Fe yields in the low intensity magnetic separation and especially in the washing of the hematite sands in the coal ash widely [12]. In the high load density and high ash coals Humphrey spirals and magnetic separation was specific performances of concentration sellectively at 82% during the washing was observed to increase the amount of high grade iron ore concentrates.

2. Flotation-magnetic separation of low grade limonite ores

In this investigation, a comparison was done between the use of poor limonite and hematite ores containing 15% limonite of the regional coal boiler bottom ash discarded as municipal waste 120 thousand tons per annum. The column flotation tests of 44% hematite using collector making hydrophobic by oleic acid and frothing agent of pine oil for long height stable frother to float limonite and hematite slimes at there stages scavangening flotation. The scavangering column flotation was nedded for long period of flotation of hematite slimes and losing high slime content at high grade froth performances [12–16]. The amount of collector and pH changed from 2 to 5 kg/t and from 8 to 11, respectively. The test results showed reduced limonite and hematite recoveries of 63% and 45%. The less amount scavangener yield resulted better (70–85%) limonite recoveries.

In the tests of reduced iron ore production reductive roasting at 1000°C with coal fine for different times, ranging 40 min, and 80 min for the limonite pellets [17–23] were converted to reduced iron ore and the properties of the reduced iron ore showed the optimum conditions of concentration. The burned asphaltite coal in reductive roasting showed that high volatile gases amount and flue gas quantity and temperature provided by microwave affected on reductive roasting of limonite slimes to direct reduced iron ore.

The flotation routes of iron ore can be classified into five major groups, i.e. cationic flotation of iron oxide, cationic flotation of quartz, anionic flotation of iron oxide, anionic flotation of quartz [24–31]. Despite the variety of flotation route for iron ores, currently, the reverse cationic flotation route widely used in the iron ore industry. The two anionic flotation methods developed by Hanna Mining and Cyanamid, i.e. direct anionic flotation and reverse anionic flotation routes, are also being used in the iron ore industry.

Tosun discussed that at the three stages of scavangering hematite slime flotation, the flotation rate of the slime suspension at 10% solid weight rate shows very low viscosity. The wash water has little effect on the froth flow. Consequently, the only froth length at 30 cm scale in the column of 1.5 m will be sufficient to 4 cm diameter column scale. Following Floatex density separation, this Outec air sparged column flotation as shown in **Figure 1** is similar to froth happens in the free fall of an 100 micron slime hematite in a positive base flow [12]. The scavangering froth time increases approximately by growing particles load of froth heavily with time at the early stages of the scavengering process. After the theree stage scavangering, The Outotec flotation column [32] use need some relatively long time, the maximum concentration of the air sparging decreases and the positive base flow effects high grade froth making. The collector cations are either stuck onto the bubble. The suspension of limonite particles caused hard water [33, 34].

Although the reverse cationic flotation route has become the most popular flotation route in iron ore industry, the direct flotation of iron oxides still appears desirable for some low grade iron ores that contain a vast amount of quartz. Oleic acids

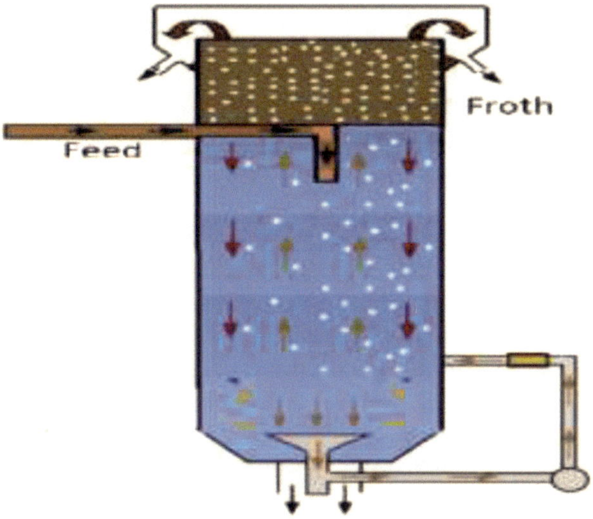

Figure 1.
Outec column flotation equipment.

as fatty acids provided high froth limonte loadseven at low dosage in the range of 0.45–0.67 kg/t [34, 35]. Reagent conditioning was affected significantly the direct froth load. The longer conditioning time can reduce reagent consumption below 50%. High conditioning time periods over 20 minutes was also found beneficial to the direct hematite flotation [36–44].

The adsorption of fatty acids on limonite plays a key role in the direct flotation route. In the literature, it is generally accepted that fatty acids adsorb on the surfaces of limonite through chemical bonding. Based on infrared studies, established that oleic acid/sodium oleate chemisorbs on limonite [39, 40]. Using the technique of micro electrophoresis, demonstrated the chemisorptions of oleic acid and lauric acid on hematite [40]. It is also confirmed chemisorption of lauric acid on hematite surfaces [44]. In addition to chemisorption, fatty acids can also adsorb on mineral surfaces through surface precipitation [37, 38].

The collector as hydroxamates that shows similar effect to fatty acids in solution [39] is used successfully in the laboratory as collectors for hematite, limonite and goethite flotation, with better performance than fatty acids [40, 41]. The adsorption mechanism of hydroxamates on limonite was classified as classical chemisorptions [35, 36].

In the reverse cationic flotation, the depressant of iron oxides that is widely used is typically corn starch. Corn starch is not soluble in cold water and must be put into solution in a process known as gelatinization.

2.1 Washing with the column flotation

Column flotation of iron ore is prefered as well yield preparation sellective floated in the microbubbles [34, 43]. Microbubble froth washing in the form of foam zones may be possible to obtain cleaner product [24, 25]. Particularly, for difficult washable vertical column is a method used successfully in flotation at high rates [40]. Particle size and type of coal as the flotation column can easily affect efficiency. However, operating parameters, especially the foam height of the column unit, the wash water is added, and the bias ratio is flammable operating parameters affect efficiency [34–39].

Other froth principles laid cyclonic column flotation cell (S-FCMC) provided a foam zone comprising inclined channels (FCMC) it proved to be effective in column frothing used. The froth product in the column has a third zone of the less froth sediment removed [37].

The application of column cells in the mineral processing industry has gone from virtually zero in 1983 to wide acceptance in 1990 [35–37]. The major operating difference between column flotation cells and mechanical flotation cells is the lack of agitation in column flotation which reduces energy and maintenance costs [34]. The practice of froth washing in direct flotation increases concentrate grades without significant recovery losses [34]. In the reverse flotation of iron ores, froth washing was found effective in reducing the loss of fine iron oxide particles to froth. It was reported that the cost of installing a column flotation circuit is approximately 25% - 40% less than an equivalent flotation circuit of mechanical flotation cells [36].

2.2 Magnetic separation Folowing reduction

Reduction in retort furnace and Shaft furnaces were commonly used in iron ore reduction processes depend on numerous factors including coal rank in carbonization, the volatile gaseous matter of coal such as presence of hydrogen, carbonyl gas and reduction rate [45–53]. Hydrate and carbon dioxide removal was stabilizing the mass desorbance, the settings of optimal diffusion conditions including structure defects (nitrogen, phosphorus, sulfur, ete.). The temperature, oxygen content of coal, optimization of carbonmonoxide concentration ratios acted the adsorption–desorption balance, the residence time and the spatial distribution of molecules in iron ore pores among other factors determining the efficiency of reduction. as factors affecting the rate and extent of char to CO motion much dependent on the site activation, its desorption properties and ore porosity.

The limonite reducing capacity of the microwave heated [54–62] column samples according to output reduced iron sand, char shale fine washed away and time sequential experiments and reduction limits were high, but high clay contents in limonite sand provided low performans washing and clay have efficiently reduced Fe yield. Samples for this microwave heat treatment at low temperatures at 900°C provided certain properties of the hematite material of high reducibility capacity and not dispersed in the wet state [63–83].

3. Material and methods

3.1 Gravity washing and floatation

The bottom ash of Avgamasya vein asphaltites represent approximately 67% of the production is carried out from the coal mines has been reduced to 120 kg sample cone reduced by up to 18 mm-fours under the hammer. Nuts are widely washed coal ash and high sulfur coal to be sold as industrial fuel asphaltites is intended to be sold. Optimum bottom ash flotation plant is determined by standard testing results performed. In the experiments, the bottom ash of Avgamasya vein asphaltites was crushed and screened prior to represent flotation samples and distribution of fractional ash is given in **Table 4**. **Figure 2** describes **Table 1** hematite distrubition versus ash size distribution. Higher sized ash contained more hematite. The sand size had lower content of hematite. Especially the range below 10 mm contained remnant about 21,3% hematite of total ash feed. The limonite percentage was 21,3% in the total bottom ash distribution. It showed uniform distribution of the hematite content in all fractions.

Screen Size	Ash Weight,%	Hematite,%	Hematite Yield,%	Limonite,%	Limonite Yield,%
+ 10	10,44	13,44	1,40	38	26
+ 5	4,8	14,6	0,70	16	22
+ 3	1,23	11,23	0,13	3	23
+ 1.8	5,54	15,54	0,86	2	24
+ 1	7,75	17,75	1,37	1	26
+0,6	13,82	13,82	1,90	2	30
+0,3	15,55	15,55	2,41	4	28
+0,1	13,74	16,74	2,30	9	32
−0,1	27,13	17,13	4,64	25	31
Total	100		15,75		27,16

Table 4.
The distribution of hematite at bottom ash regarding particle size and hematite yield content in ash.

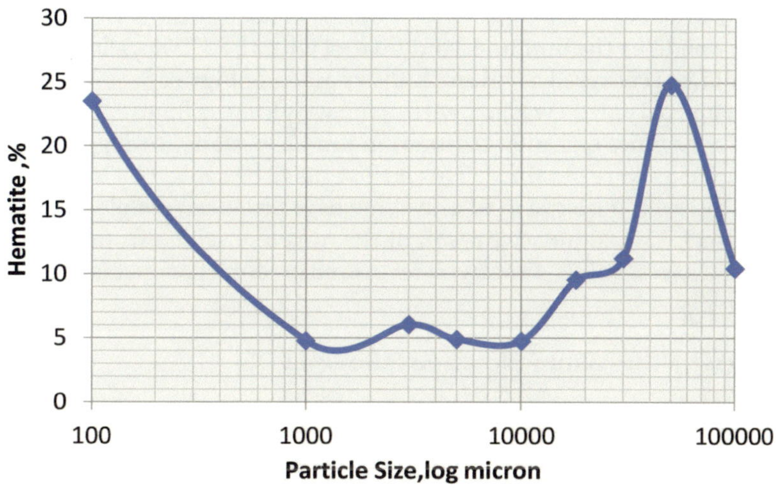

Figure 2.
Content distributiobn values in hematite yields in different size grain fraction against size.

3.2 Flotation of hematite/limonite in oleate

30 kg samples were ground to −0,1 mm in ball mill and 1 kg representative samples were used in the study represent −0,3 mm size of −0,1 mm separated flotation tests for grain size fractions were subjecte direct flotation made by oleic acid. In the direct flotation test; oleic acid at neutral pH solution were used as collector. – 0,1 mm fraction used in this study were studied in similar way.

1 liter Denver laboratory flotation cell for clean conventional flotation tests were used to produce concentrated hematite and limonite concentrates. The conditioning 5 min and frothing 10 min in conventional flotation carried for 2 min extra at 20% weight solids. The flotation cell was agitated in a mixing speed 1500 rpm. Limonite flotation tests used oleic acid 1000 g /ton and frother pine oil 400 g /t to be conditioned.

According to the results of the flotation made on pH effect on Limonite at size classed; −0,1 mm grain obtained by gravity concentrated classes test results as given in **Figure 3**.

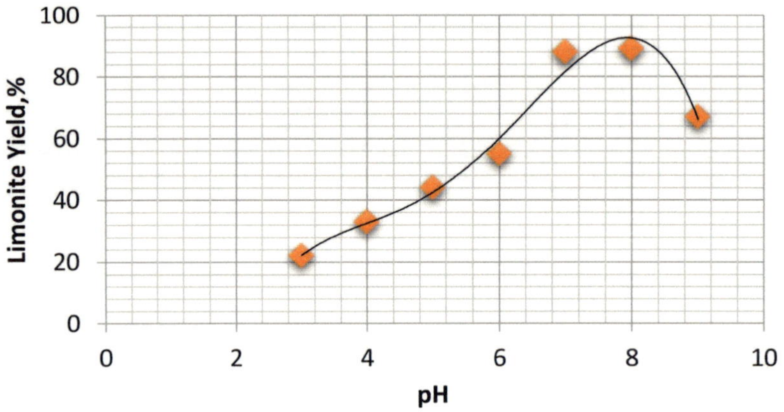

Figure 3.
Test values in limonite yields in- 0.3 mm grain fraction against pH.

Limonite yields were illustrated in **Figure 1** with the curve.

Limonite concentrate can be floated in a weight ratio of 58.2%. 59.7% iron grade limonite can be floated in a weight ratio of 68.2% concentrations at pH 8 (**Figure 3**).

26.3% by weight of the limonite sand floated with the 57.5% yield of limonite. The sand could be recovered as given 28.4% as given in **Figure 5**

Crusher-run coal can be washed with some weighing as high as 17.9% when the limonite and ash slimes on limonite flotation for 42.3% constituting 0.02 mm grain size flotation yields were obtained. This is thought to be caused by iron ore slime Fe content. However, limonite sand and slime have also been coupled in parallel as limonite product. The cumulative result of the mixed obtained from the test; 76.5% side with an efficiency of 28.4% can be recovered as slime product is seen from **Figure 4**.

3.3 Washing with the column flotation followed FDS

The Floatex Density Separator (FDS) is a water sparged-bed gravity separator which is used to separate different density of hematite ores at heavier specific gravity. Both the size below 0,5 mm and density have substantial effective on the concentration of limonite and hematite. The FDS was attracted considerable interest in iron ore concentration. The macropictures of samples are shown in **Figures 5** and **6**.

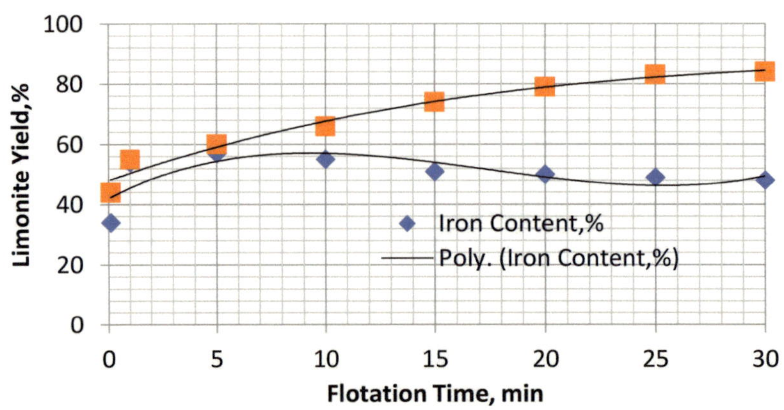

Figure 4.
Test values in limonite yields in- 0.3 mm grain fraction against flotation time.

Figure 5.
The macropictures of Şırnak limonite.

Figure 6.
The macropictures of Şırnak hematite.

The separator consists of an upper tank with a square cross-section and a lower conical section [30]. Representative of −0.5 mm samples are concentrated and then concentrate hematite of FDS reduced to grinding −100 mikro at controlled size grinding. The representative model, 1,6 m glass column 3 cm in diameter laboratory column cell flotation cell (**Figure 1**) used in the column flotation unit of limonite slime flotation.

The reagents used in conventional flotation column flotation tests were also performed in the column tests. In the column flotation tests oleic acid 1000 g / ton pine oil 400 g / t were conditioned foam height is kept constant at 30 cm. Zero Bias ratio is used to concentrate hematite ore and limonite sands. The flotation time was used for 3 min and 35 min time condition coal in tests. 10% solid/liquid ratios of 200 ml/min the wash water rate were used in the experimentation.

Column flotation tests results from limonite concentrate, shale waste can be taken as sink bottom product and limonite yield equilibrium distribution is given in **Table 5**. Accordingly (−100microns) mm grain size in bottom ash is mixed with slime limonite can be as 60.60% in cumulative yield will be thrown when recovered limonite is contained, 54.3% iron content of the ash and bottom tailings contained 4% iron as waste (**Figure 7**).

Column Flotation efficiency of limonite products produced from the results of tests of the direct flotation to 77% of limonite yield has fallen 64% value.

% Component	Şırnak Kızılsu Hematite Sand	Şırnak Limonite	Şırnak Şenoba Limonite
SiO_2	3,53	4	11,14
Al2O3	2,23	6,5	8,61
Fe2O3	49,1	53	54,9
CaO	23,48	5,23	9,18
MgO	2,2	2,18	4,68
K_2O	0,41	0,53	3,32
Na_2O	0,35	0,24	1,11
Ignition Loss	16,19	26,11	38.43
SO3	0,32	0,21	0,2

Table 5.
The chemical analysis values of various limonite ores, in calcareous formations of Şırnak province.

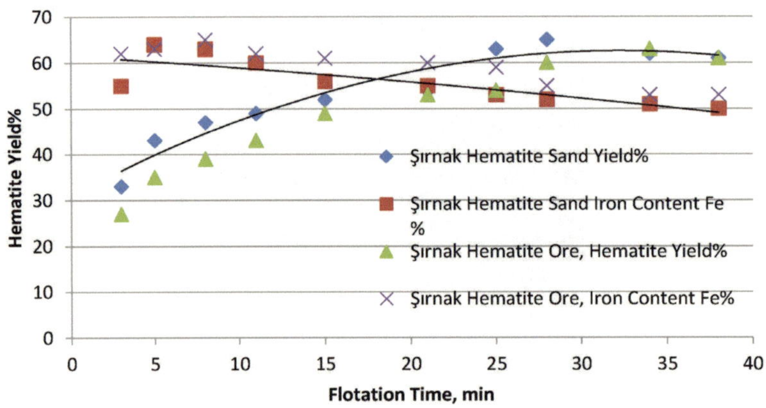

Figure 7.
Hematite yields in- 0.1mm grain fraction column flotation tests of Şırnak Hematite Sand and Ore at pH 8.

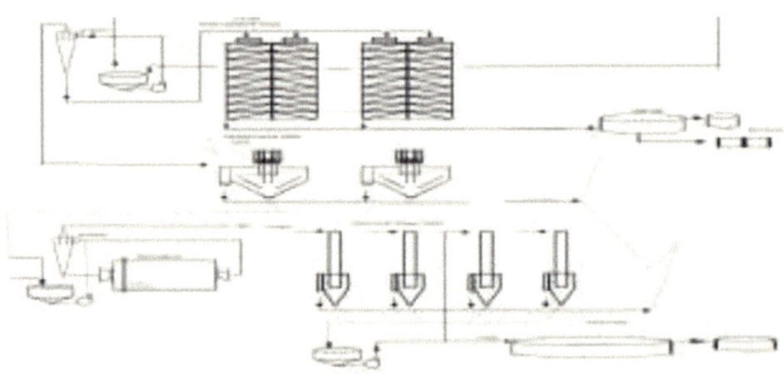

Figure 8.
Proposed Limonite concentration plant.

Flotation test results produced for the limonite product yields from 77–64% of the iron content has fallen 45% value (**Figure 8**). As shown in **Figure 5** The iron contents produced from test results of 42% decreased to value of 37% for the column flotation of bottom ash. The product yield is lower compared to other methods.

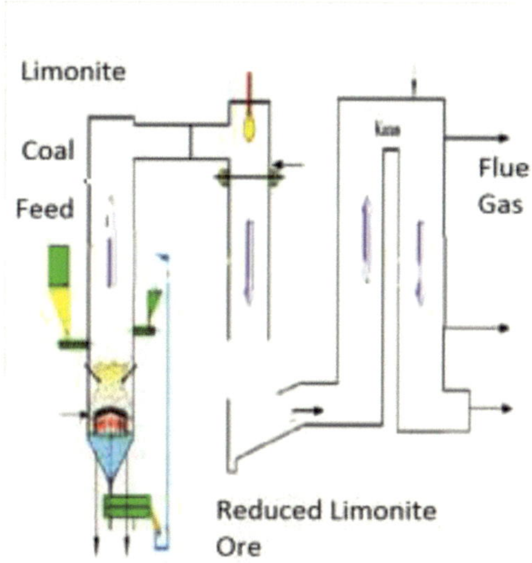

Figure 9.
Schematic view of vertical shaft bubling reduction with recycled by microwave heating technique.

3.4 Reducing bubbling Teatment

Reduction was carried out in 5000 ml tubes by adding 75 gr limonite to 1900 ml of bubling bath. For a homogeneous bubling suspension limonite coal asphaltite mixture was first subjected to cold start treatment in a 5 minutes microwave radiation (**Figure 9**).

After the hot temperature gas limonite suspension was allowed to stand for 30 minutes after being bubleded so that the reduced particles and ashes were collected and settled cyclone output. At the end of the period, suspended limonite concentrate was reduced by cycled method and was magnetically separated from ashed matter fine.

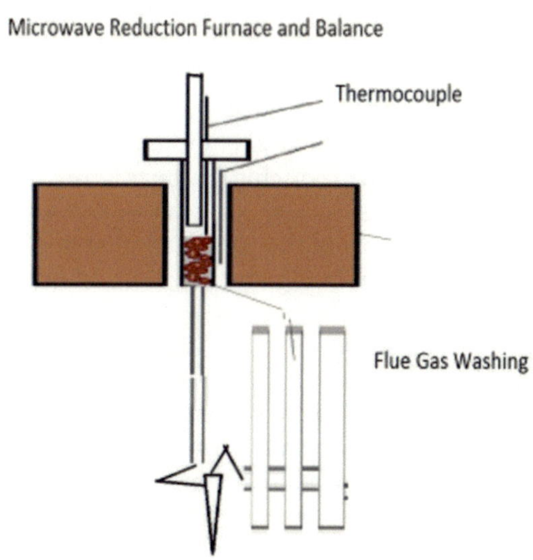

Figure 10.
TGA analyzes of the samples were carried out in Bubling bed in microwave radiated tube.

The layout of the reducing cycle was somewhat simpler than that of the limonite slurry: there was no gas reducing column towers connected to the ashed contaminated bubbling compost, and the reducing unit contained one single microwave radiation column can be used to perform the three separation magnetized flow phases: roughing, scraping and cleaning. The variation of the third cycle washed was also limited recycled by microwave act (**Figure 10**).

4. Results and discussion

The current use of absorbent hematite and new areas of use increase in demand due to outflow. Absorbant matter beds must be fully identified, potential sources should be determined, absorbant purpose on fulness should be investigated and suitable properties for absorbent compost production. This study was improved reduction in microwave processes. This study used for microwave reduction heat absorbant hematite samples taken from the local area. By applying the reductive atmosphere processes, The standard semimetallic materials was suitable for industrial microwave radiation emitted through hematite for reduced matter production was possible.

4.1 Bubling reducing activation

4.1.1 Coal gas and bath sand particle size

A major reason is that the retention time in fixed film processes is longer than in solid–gas processes. This allows more time to the carbonization far cracking to the desorbed persistent compounds. Furthermore, high rank coals allows an sufficient intimate contact between surface pores and gas atmosphere in the furnace due to more gas desorptions [55–56].

4.1.2 Bubling bed porosity

The bubling fluid matter was simulated for reduction solid air mixtures on sands. The sand/air mixture was more than occasionally exposed to limonite sand to air. The air fluid was comprised of a unique high-stability base plus high-performance oxidation inhibitor/stabilizer (**Figure 10**).

Initially, most of the limonite flow occured through chemical reduction of the ıron oxides over the hematite where the coal combustion temperature was in the combustion phase below 950°C that reducing gas lasts approximately in 5–13 mins. The gas removal efficiency from asphaltite coal was 40–90% reported during the temperature range at 950°C. Following the combustion at 800°C iron ore reductions over hematite was started as shown in **Figure 9**.

4.2 Bubling bed reducing treatment

Bubbling reduction design of fluidized bed made an advantage of reducing ore and eliminating the effects of microwave convection. The balance and gas flow measuring ensuring the heat power and temperature measurement is made in a magnetic field of known cycled flow and that the heat losses in the system minimized under both magnetic heating [67–83].

We report on this study of the hematite and limonite compound in terms of iron oxide weight stability. Microwave thermal stability in column layers radiated thermal activation determined 800°C temperature value of the limonite and hematite in

Material	The Chemical composition	The Bulk density g/cm^3	Microwave Activity	Reduction Rate at 30 min CO flow
Ferrite pellet	CaO:2,1–2.8%; SiO2 ≤ 10%;Fe2O3:45 ~ 65%; P ≤ 0.1%;S ≤ 0.1%,	2.8 ~ 3.2	Weak	75%
The sintered ore	CaO:1.16%; SiO2:5.18%;Total Fe:54.3%;P:0.05%; S:0.02%	2.4 ~ 3.0	Strong	85%
The iron oxide powder	total iron is 55%, Oxide	4.8 ~ 5.2	Strong	65%
Limonite slime, oxide	total iron oxide 70.02%, silica 1.55% .	3.6 ~ 4.3	Weak	65%

Table 6.
Microwave activity, chemical composition and physical properties of iron materials in the experiments.

this chemical compound in a crystal structure of diamagnetic grains can be emitted at an optimum temperature of 980°C.

The microwave reduced chemical composition and physical properties of FeO formation material in reduced matter showed as **Table 6**. The bulk density average value of the ferrite pellet is 3.02 g/cm^3 by reduction. In order to better application of the hematite ferrite pellet the melting point of the ferrite pellet and slag formation material of reducer was compared. The finally the hematite sample put into microwave furnace at the temperature of 1350°C and time of holding temperature was 2 minutes. The reduced sintering state of sample was immediately observed after cooling.

4.3 Concentration reduced ore and model method

The iron oxide groups showed that the limonite quality will enhance the quality of direct reduced iron ore quality as high iron grade. Further reductive roasting provides much higher metal conversion. The concentration model using Retort furnace and coal char was seen in **Figure 11**.

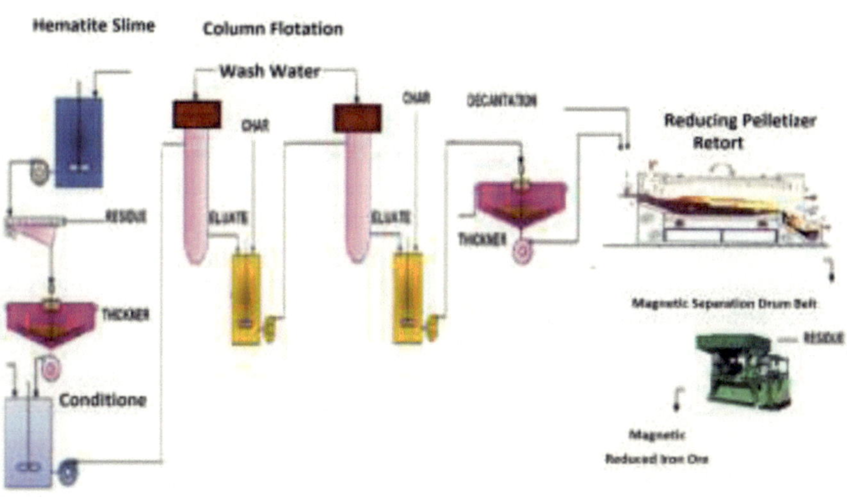

Figure 11.
Concentration and reduced iron ore production by retort pelletizing bed in retort furnace.

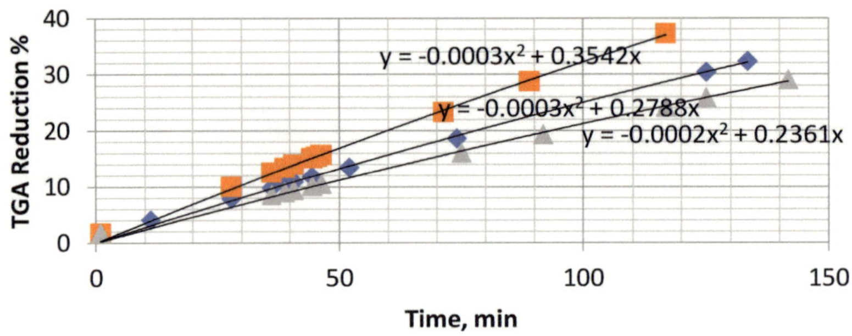

Figure 12.
TGA analyzes of the samples were carried out in Bubling bed in microwave radiated tube.

4.4 TGA analysis

Analyzes were performed using approximately 300 g of limonite samples at a temperature range of 800–1000° C at a heating rate of 10° C / min. The temperature ranges and mass loss values obtained from the TG analysis are given in **Figure 12**. The limonite samples reduced 65% iron yield at 805°C, 899°C and 980°C, respectively. The limonite reduction was effective due to the core heat adsorbed on autonomous heating particle core by radiation.

In the TGA curves, the reduction rate values of the microwave activated limonite samples were determined to be lower than those of the native sample due to the increasing iron ore contents at 72–81%. The hematite, coal char shale and limonite samples revealed that the total mass losses at 1000°C were 31.9%, 47.75% and 38.15%, respectively.

5. Conclusions

In the pH measurements made, the pH value of 7,3 in washing limonite finally at the last washing column flotation decreased to iron yield lower to 53%, depending on the concentration of salt content of slurries in the foam water.

In the three stage microwave cycling reduction test measurements made with found that 730 g Fe/kg in limonite/asphaltite char decreased to 530 g Fe /kg in last column output. Likewise, the reduced limonite ore iron grade increased to 76% Fe as final magnetic separation obtained after 10 min concentrating by low permanent magnet and Fe yield was at 47% of final performance.

The reduced iron ore pellets or sinters provided high quality steelmaking feeds in plant facilities. The other side in steelmaking is low impruty scrap feed and reduced iron contents. In our country, Iskenderun EAF needed this quality ironmaking reduced iron ores in steel ladles. There is lack of local miners to produce direct reduced iron ore and fluxing limed pellets which are working at ironmaking. The microwave reduction method use in this study proved that investing in high quality ironmaking and steelmaking feed.

Due to the low iron content of Şırnak limonites and hematite sands this effective method showed that retort reduction and magnetic separation following column flotation might be effective as bubble bed reduction. The total iron yield was also determined as 72% at not reduced sufficiently, as well as limonite product can be 56%. The recovery of limonite was 87% in the limonite flotation plant. The recovery in column flotation can be reduced to 68% of the bottom ash disposed.

Limonite content of the bottom ash was suitable for evaluation in sponge iron production so that the sponge product will provide benefits in terms of reduced costs as well as transport and environmental protection. The lower grade value can be both beneficial in recovery and capacity for industrial iron produced.

In the proposed design of the microwave direct reduction following limonite flotation plant, reduced iron ore product was capable of total iron grade of 81% in the plant, as reduced iron (12–10 mm) 350,000 tons, industrial 80% iron grade may produced with microwave radiation and char use (−10 / -0.5 mm) of 20,000 tons, as the magnetic separated 165,000 tons of reduced iron was able to be produced in proposed model in **Figure 8**.

Iron ores fines as limonite and hematite sands slime was finely concentrated by column flotation then managed efficient reduction by bubbling bath at 25% porosity at temperature of reducing at 900°C. It was determined that so great extent reducing rate 64% at 20 minutes and 72% of iron recoveries was found by a porous bubbling bath over 47–55% weight rate of limonite was optimized for reduction. The more efficiently conducted heat reduced more a low amount limonite with below iron weight rate of 35%.

This research was included primitive microwave ovens, but that smaller tubing flows reduction in microwave as sintering tube plants without any complicated equipment for operation. Two principle stages could still manage prospective pre reduction granule and pellet production in new sintering plants: improvement of preparation of limonite ores of minus 5 mm retort reduction process as well as optimization of the retort coal gas reduction quality. A second stage was principle microwave reduction tube passage has prevailed during the last reduction because of hematite core was becoming higher temperature CO reduction due to the act of the microwave radiation over core of particles This process was beneficial for using low quality coal gas and not needing sintering on the issue of energy and environmental dust emission ways.,Due to the produced granules was efficiently used in electrical arc furnaces for scrape steel and other mixture demands, in the steel plant operators improve scrape and dust controls in hot metal pouring. Approx. 5% of the total pig iron production was originated from the pre reduced iron ores ores of total 35 milliontons of pig iron production in EAF technology, which highly struggle on recover waste gas energy and dust, toxic emissions. This technology may imrove dust control by stick hot matter in hot pouring process.

Author details

Yildirim İsmail Tosun
Engineering Faculty, Mining Engineering Department, Şirnak University, Şirnak, Turkey

*Address all correspondence to: yildirimismailtosun@gmail.com

References

[1] World Steel Association, 2020, 2020 World Steel in Figure, Statistics http://www.worldsteel.org

[2] World Steel Association, 2019, Steel Statistics http://www.worldsteel.org

[3] TÇÜD, 2020, Demir Çelik Üretim İstatistikleri

[4] Hazal Kermen, 2019, Bölüm 1: Demir-Çelik Sektörü. Türkiye Çelik Üreticileri Derneği, 2018, On birinci kalkınma planı (2019-2023) ana metal sanayii çalışma grubu raporu,T.C.Kalkınma Bakanlığı, Ankara.

[5] Yıldız, N., "Demir Cevheri", 2010, "Cevher üretimi, zenginleştirilmesi, peletlenmesi, sinter üretimi, sünger demir üretimi, çelik üretimi", Ertem Basım Yayın Dağıtım San. s 248, ISBN 978-975-96779-3-0,Ankara.

[6] SIMA, 2020, Sponge Iron/DRI, http://www.spongeironindia.com/statistics.php

[7] SIMA, 2017, Sponge Iron Manufacturers Association, New Delhi - DRI Update, http://www.spongeironindia.com/ images/ publications/ DRI%20UPDATE%20 -%20December%202017.pdf

[8] Anonymous, 2020, Can 10% Royalty Reduction In Iron Ore Give Fillip To Atmanirbhar Bharat?, **Outlook Magazine**, https://www.outlookindia.com/website/story/india-news-can-10-royalty-reduction-in-iron-ore-give-fillip-to-atmanirbhar-bharat/360757

[9] Anonymous, 2020, Indian iron ore pellet makers raise supply concerns, **Argus Media**, https://www.argusmedia.com/en/ news/ 2133647-indian-iron-ore-pellet-makers-raise-supply-concerns,

[10] Tosun Y.I., Recovery Of Hematite from The Asphaltite Boiler's Bottom Ash By Column Flotation – Plant Modelling, EJONS International Journalon Mathematic, Engineering and Natural Sciences, Year:2 Volume: (2) Publication Date: June 1, 2018 , ISSN 2602 4136, www.ejons.co.uk, pp102-115

[11] Tosun, Y.I., Rowson, N.A., Veasey, T.J., 1994, "The Bio-column Flotation of Coal for Desulphurization and Comparison with Conventional and Column Flotation", 5th Int. Conf. on Mineral Processing, Nevsehir.

[12] Tosun, 2016, Recovery of Hematite from The Coal Boiler's Bottom Ash by Column Flotation for Sponge Iron Production, SWEMP 2016, 16th Internatıonal Symposıum on Envıronmental Issues and Waste Management ın Energy and Mıneral Productıon, October 5-7, Istanbul, Turkey

[13] IEA, 2007, IEA Coal Research Ltd, Clean Coal Technology Report, (A.J. Minchener and J.T. McMullan)

[14] Krishnan, S.V., Iwasaki, I., 1984. Pulp dispersion in selective desliming of iron ores. Int. J. Miner. Process. 12, 1-13.

[15] DeVaney, F.D., 1985. Iron Ore. In: Weiss, N.L. (Ed.), SME Mineral Processing Handbook, American Institute of Mining, Metallurgical and Petroleum Engineers, New York.

[16] Yang, D.C., 1988. Reagents in Iron Ore Processing. In: Somasundaran, P. and Moudgil, B.M., (Eds.), Reagents in Mineral Technology, Marcel Dekker Inc., New York, pp. 579-644.

[17] Peck, A.S., Raby, L.H., Wadsworth, M.E., 1966. An infrared study of the flotation of limonite with oleic acid and sodium oleate. Trans. AIME 235, 301-307.

[18] Anonymous b, 2015, Yakma Kazanları, http://**www.alfakazan.com.tr**, Alfa Kazan ve Makine AŞ,Ankara

[19] Anonymous c, 2015, Kalina Cycle, http://**www.imparator enerji.com.tr**, İmparator Enerji, GeoPower, İstanbul

[20] Anonymous d, 2015, Akışkan Yataklı Yakma Kazanı, http://**www.mimsan.com.tr**, Mimsan A.Ş. , İstanbul

[21] Anonymous e, 2015,http://www.atlasinc.dk/

[22] Anonymous f, 2015,http://www.santes.com.tr/

[23] Anonymous g, 2015, http://www.ottusa.com/mobile_ systems/ mobile_systems.htm

[24] Fuerstenau, M.C., Miller, J.D., Gutierrez, G., 1967. Selective flotation of iron ore. Trans. AIME 238, 200-203.

[25] Fuerstenau, M.C., Harper, R.W., Miller, J.D., 1970. Hydroxamate vs. fatty acid flotation of iron oxide. Trans. AIME 247, 69-73.

[26] Raghavan, S., Fuerstenau, D.W., 1974. The adsorption of aqueous octylhydroxamte on ferric oxide. J. Colloid Interface Sci. 50, 319-330.

[27] Papini, R.M., Brandão, P.R.G., Peres, A.E.C., 2001. Cationic flotation of iron ores: amine characterisation and performance. Minerals & Metallurgical Processing 17, 1-5.

[28] Vidyadhar, A., Hanumantha Rao, K., Chernyshova, I.V., Pradip, Forssberg, K.S.E., 2002. Mechanisms of amine–quartz interaction in the absence and presence of alcohols studied by spectroscopic methods. J. Colloid Interface Sci. 256, 59-72.

[29] Dobby, G., 2002. Column Flotation. In: Mular, A.L., Halbe, D.N. and Barratt, D.J. (Eds.), Mineral Processing Plant Design, Practice and Control, SME Inc., Littleton.

[30] Peres, A.E.C., Araujo, A.C., Shall, H., Zhang, P., Abdel-Khalek, N.A., 2007. Plant practice: nonsulfide minerals. In: Fuerstenau, M.C., Jameson, G., Yoon, R-H (Eds.), Froth Flotation – A Century of Innovation, SME, Littleton.

[31] Anonymous, 2015, *Multotec Şirket* web sayfası, http://www.multotec.com/category/ industry/coal

[32] Anonymous, 2015, Outotec co column floatation, https://www.outotec.com/products-and-services/newsletters/minerva/minerva-issue-2--2017/flotation-columns-getting-the-most-from-fine-ores/

[33] Han, K.N., Healy, T.W., Fuerstenau, D.W., 1973. The mechanism of adsorption of fatty acids and other surfactants at the oxide-water interface. J. Colloid Interface Sci. 44, 407-414.

[34] Buckland, A.D., Rochester, C.H., Topham, S.A., 1980. Infrared study of the adsorption of carboxylic acids on limonite and goethite immersed in carbon tetrachloride. Faraday Trans. I 76, 302-313.

[35] Somasundaran, P., Huang, L., 2000. Adsorption and aggregation of surfactants and their mixtures at solid/liquid interfaces. Adv. Colloid Interface Sci. 88, 179-208.

[36] Zhang, R., Somasundaran, P., 2006. Advances in adsorption of surfactants and their mixtures at solid/solution interfaces. Adv. Colloid Interface Sci.123, 213-229.

[37] Bulatovic, S.M., 2007. Handbook of Flotation Reagents, Elsevier, Amsterdam.

[38] Finch J.A., Dobby,G.S.(Eds), 1990, *Column Flotation*, Pergamon Press, Toronto

[39] Hadler, K., M. Greyling, N. Plint, and J. J. Cilliers. 2012. The effect of froth depth on air recovery and flotation performance. *Minerals Engineering* 36: 248-253.

[40] Jameson , G. J. 2001. The flotation of coarse and ultrafine particles. *International Journal of Mineral Processing* 72 : 12-15

[41] Yoon, R. H. 1993. Microbubble flotation. Minerals Engineering 6(6): 619-630.

[42] Yoon, R. H. 2000. The role of hydrodynamic and surface forces in bubble–particle interaction. *International Journal of Mineral Processing* 58(1): 129-143.

[43] Yianatos, J. B., J. A. Finch, and A. R. Laplante. 1988. Selectivity in column flotation froths. *International Journal of Mineral Processing* 23(3): 279-292.

[44] Rubio, J. 1996. Modified column flotation of mineral particles. *International Journal of Mineral Processing* 48(3): 183-196

[45] J. P. Hansen, N. B. Melcher & M. M. Fine, 1961,Pre-Reduced Iron-ore Pellets, their Experimental Preparation, *JOM* volume13, pages 314-315(1961)

[46] Fine, M. M.; Melcher, N. B.; Bernstein, N.; Woolf, P. L. & Reuss, J. L., 1970. Prereduced Iron Ore Pellets: Preparation, Properties, Utilization, United States Bureau of Mines Reports, USBM Bulletin 651,

[47] Melcher ,N. B., 1963, Smelting Prereduced Iron Ore Pellets, *JOM* volume15, pages298-301(1963)

[48] Machida S, Sato H, Takeda K, 2009, Development of the Process for Producing Pre-reduced Agglomerates, *JFE GIHO* No. 22, p. 25-31

[49] M. Kumar ; S. K. Patel , 2009, Characteristics of indian non-coking coals and iron ore reduction by their chars for directly reduced iron production, Mineral Processing and Extractive Metallurgy Review, 28,3,258-273

[50] T.R. Ramachandra Rao , 2006, Direct Reduced Iron Industry in India — Problems and Prospects, *Proceedings of the International Seminar on Mineral Processing Technology, Chennai, India. pp. 461-463.*

[51] *J.E. Rehder,1983,* Manufacturing cast iron with pre-reduced iron ore pellets, *US Patent no 4401463*

[52] M. Small, 1981, Direct Reduction of Iron Ore, JOM, Volume 33, issue 4, p 71-75

[53] Subhasisa N , 2009, Study of Reduction kinetics of Iron ore Pellets by Noncoking coal, MSc Thesis, Department of metallurgical and materials engineering, national institute of technology, Rourkela, India

[54] Tosun, 2016, Microwave Roasting of Pyrite and Pyrite Ash for Sponge Iron Production, SWEMP 2016, 16th International Symposium on Environmental Issues and Waste Management in Energy and Mineral Production, October 5-7, Istanbul, Turkey

[55] Amankwah, R.K., Pickles, C.A., 2005. Microwave calcination and sintering of manganese carbonate ore. Canadian Metallurgical Quarterly 44 (2), 239-248.

[56] Amankwah, R.K., Pickles, C.A., Yen, W.T., 2005. Gold recovery by microwave augmented ashing of waste activated carbon. Minerals Engineering 18 (2), 517-526.

[57] Chen TT, Dutrizac JE, Haque KE, Wyslouzil W, Kashyap S. 1984, The relative transparency of minerals to microwave radiation. Canadian

Metallurgical Quarterly, 123, 3, s. 349-51.

[58] Datta A K; Nelson S O (2000). Fundamental Physical Aspects of Microwave Absorption and Heating in Handbook of Microwave Technology for Food Applications. CHIPS Publications, USA

[59] Datta A K; Sun E; Solis A (1995). Food dielectric property data and their composition-based prediction. In: Engineering Properties of Foods (Rao M A; Rizvi S S, eds), Chapter 9, 457-494. Marcel Dekker, Inc., New York

[60] Decareau R V (1985). Microwaves in the Food Processing Industry. Academic Press, Orlando, FL, USA

[61] El-Shami S M; Selim I Z; El-Anwar I M; Hassan M M (1992). Dielectric properties for monitoring the quality of heated oils. Journal of the American Oil Chemists' Society (JAOCS), 69(9), 872-875

[62] Gabriel C., Gabriel S., Grant E.H., Halstead B.S.J., Mingos D.M.P., 1998, Dielectric parameters relevant to microwave dielectric heating. Chemical Society Reviews, 27, s.213-23.

[63] Haque KE. Microwave energy for mineral treatment processes—a brief review, 1999, International Journal of Mineral Processing, 57, 1, s.1-24.

[64] Hutcheon, R.M., De Jong, M.S., Adams, F.P., 1992. A system for rapid measurement of RF and microwave properties up to 1400 _C. Journal of Microwave Power and Electromagnetic Energy 27 (2), 87-92.

[65] Hutcheon, R.M., De Jong, M.S., Adams, F.P., Lucuta, P.G., McGregor, J.E., Bahen, L., 1992a. RF and microwave dielectric measurements to 1400 _C and dielectric loss mechanisms. In: Materials Research Society Symposium Proceedings (Microwave Processing of Materials III), vol. 269, pp. 541-551.

[66] Hutcheon, R.M., Hayward, P., Smith, B.H., Alexander, S.B., 1995. High-temperature dielectric constant measurement – another analytical tool for ceramic studies. Microwaves: Theory and Application in Materials Processing III, vol. 59. Ceramic Transactions, American Ceramic Society, pp. 235-241.

[67] Jacob J., Chia L.H.L., Boey F.Y.C.,.1995, Review—thermal and non-thermal interaction of microwave radiation with materials. Journal of Materials Science, 30, 21, s.5321-7.

[68] Karmazsin, E., 1987. Use of low – and high-power microwave energy for thermal analysis. Thermochimica Acta , 110, 289-295.

[69] 9th International Exergy, Energy and Environment Symposium (IEEES-9), May 14-17, 2017, Split, Croatia

[70] Kelly RM, Rowson NA., 1995, Microwave reduction of oxidised ilmenite concentrates. Minerals Engineering, 8, 11, s.1427-38.

[71] Kılıç Ö., 2009, Mikrodalga ile Isıl İşlem Uygulamanın Kireçtaşı Kalsinasyonuna Etkisi, Madencilik, 48, 3, s 45-53.

[72] Kingman S.W., Vorster W., Rowson N.A., 1999, The influence of mineralogy on microwave assisted grinding. Minerals Engineering, 3,3, s.313-27.

[73] Lu, T., Pickles, C.A., Kelebek, S., 2007. Microwave heating behaviour of a gibbsite type bauxite ore. In: Bekguleryuz, M.O., Paray, F., Wells, M. (Eds.), Proceedings of Symposium on Light Metals in Transport Applications. MetSoc (CIM), Toronto, Ont. Canada, pp. 421-449 (August 25-30).

[74] Ma, J., Pickles, C.A., 2003. Microwave segregation process for nickeliferous silicate laterites. Canadian Metallurgical Quarterly 42 (3), 313-326.

[75] Marland S, Han B, Merchant A, Rowson N., 2000, The effect of microwave radiation on coal grindability. Fuel, 79, 11, s.1283-8.

[76] Metaxas, A.C., Meredith, R.J., 1983. Industrial Microwave Heating. Chapter 10, Peter Peregrinus, London, UK.

[77] Salsman J.B., Williamson R.L., Tolley W.K., Rice D.A., 1996, Short-pulse microwave treatment of disseminated sulphide ores. Minerals Engineering, 9, 1, s.43-54.

[78] Standish, N., Worner, H.K., Gupta, G., 1990. Temperature distribution in microwave heated iron ore–carbon composites. J. Microwave Power Electromagnet Energy 25 _2., 75-80.

[79] Standish, N., Worner, H.K., Obuchowski, D.Y., 1991. Particle size effect in microwave heating of granular materials. Powder Technology 66, 225-230.

[80] VanWyk EJ, Bradshaw SM, de Swardt JB., 1998 The dependence of microwave regeneration of activated carbonon time and temperature, Journal of Microwave Power and Electromagnetic Energy ,33, 3, s.151-7

[81] Veasey TJ, Fitzgibbon KE., 1990, Thermally assisted liberation—a review. Minerals Engineering , 3, 1/2, s.181-5.

[82] Walkiewicz JW, Kazonich G, McGill SL., 1988, Microwave heating characteristics of selected minerals and compounds. Minerals and Metallurgical Processing , 5, 1, s.39-42.

[83] Xia D.K., Pickles C.A., 2000, Microwave caustic leaching of electric arc furnace dust, Minerals Engineering, 13, 1, s.79-94.